Environment and Nuclear Energy

Environment and Nuclear Energy

Edited by

Behram N. Kursunoglu
Global Foundation, Inc.
Coral Gables, Florida

Stephan L. Mintz
Florida International University
Miami, Florida

and

Arnold Perlmutter
University of Miami
Coral Gables, Florida

Springer Science+Business Media, LLC

Library of Congress Cataloging-in-Publication Data

Environment and nuclear energy / edited by Behram N. Kursunoglu, Stephan L. Mintz and Arnold Perlmutter.
p. cm.
"Proceedings of an International Conference on Environment and Nuclear Energy, held October 27–29, 1997, in Washington, D.C.—T.P. verso.
This International Conference on Environment and Nuclear Energy was the 20th of a series of conferences on energy that were began in 1977—Preface.
Includes bibliographical references and index.
1. Nuclear energy—Congresses. 2. Environmental protection—Congresses. 3. Nuclear energy—Government policy—Congresses. I. Kurşunoğlu, Behram, 1922– . II. Mintz, Stephan L. III. Perlmutter, Arnold, 1928– . IV. International Conference on Environment and Nuclear Energy (20th: 1997: Washington, D.C.) TK9006.E54 1998
333.792' 414—dc21 98-41249
CIP

Proceedings of an International Conference on Environment and Nuclear Energy, held October 27 – 29, 1997, in Washington, D.C.

This volume was taken from a series of conferences sponsored by Global Foundation, Inc., Coral Gables, Florida

ISBN 978-1-4899-1548-1 ISBN 978-1-4899-1546-7 (eBook)
DOI 10.1007/978-1-4899-1546-7

Originally published by Plenum Press, New York in 1998
Softcover reprint of the hardcover 1st edition 1998

http://www.plenum.com

10 9 8 7 6 5 4 3 2 1

PREFACE

This International Conference on Environment and Nuclear Energy was the twentieth in a series of conferences on energy that began in 1977. The focus of the conference was on the use of energy and the impact on environment. Could nuclear energy be a major contributor to an energy source mix that would help alleviate greenhouse gas emissions and global warming problems? To what extent would nuclear energy help solve pollution problems such as acid rain resulting from heavy use of fossil fuels, particularly coal. Do the issue of nuclear waste and a possible nuclear weapons proliferation present scientific, technological, and political challenges? If all these problems inhibit the global use of nuclear energy, then the world is facing an important issue requiring global solution.

The conference examined the impact of nuclear energy on regional and global environmental issues concerning a variety of scenarios including (1) competition in deregulated energy environments; (2) constraints levied upon use of fossil energy; (3) and the potential for nuclear energy beyond the generation of electricity (e.g., process heat, fuels production, etc.)

Closely coupled with the examination of the environment and nuclear energy was the assessment and role of nuclear energy in meeting future energy needs of the growing world population and economic development. Similarly, the role of nuclear power in meeting national and regional energy security objectives was also addressed. The 1998 meeting will take place October 21-23, in Paris, and October 22-24, 1999, in Washington, D.C. concerning the "Impact of Nuclear Power on Environment."

The Chairman of the Board and the Trustees of the Global Foundation wish to acknowledge not only the financial support, but also the intellectual input of the Electric Power Research Institute.

Behram N. Kursunoglu
Stephan L. Mintz
Arnold Perlmutter
Coral Gables, Florida
February 1997

ABOUT THE GLOBAL FOUNDATION, INC.

The Global Foundation, Inc., was established in 1977 and utilizes the world's most important resource . . . people. The Foundation consists of senior men and women in science and learning; outstanding achievers and entrepreneurs from industry, governments, and international organizations; and promising, enthusiastic young people. These people form a unique and distinguished interdisciplinary entity, and the Foundation is dedicated to assembling all the resources necessary for them to work together. The distinguished senior members of the Foundation convey their expertise and accumulated experience, knowledge, and wisdom to the younger membership on important global issues and frontier problems in science.

Our work is a common effort, employing the ideas of creative thinkers with a wide range of experience and viewpoints.

GLOBAL FOUNDATION BOARD OF TRUSTEES

GLOBAL FOUNDATION'S RECENT CONFERENCE PROCEEDINGS

Making the Market Right for the Efficient Use of Energy
Edited by: Behram N. Kursunoglu
Nova Science Publishers, Inc., New York, 1992

Unified Symmetry in the Small and in the Large
Edited by: Behram N. Kursunoglu, and Arnold Perlmutter
Nova Science Publishers, Inc., New York, 1993

Unified Symmetry in the Small and in the Large - 1
Edited by: Behram N. Kursunoglu, Stephan Mintz, and Arnold Perlmutter
Plenum Press, 1994

Unified Symmetry in the Small and in the Large - 2
Edited by: Behram N. Kursunoglu, Stephan Mintz, and Arnold Perlmutter
Plenum Press, 1995

Global Energy Demand in Transition: The New Role of Electricity
Edited by: Behram N. Kursunoglu, Stephan Mintz, and Arnold Perlmutter
Plenum Press, 1996

Economics and Politics of Energy
Edited by: Behram N. Kursunoglu, Stephan Mintz, and Arnold Perlmutter
Plenum Press, 1996

Neutrino Mass, Dark Matter, Gravitational Waves, Condensation of Atoms and Monopoles, Light Cone Quantization
Edited by: Behram N. Kursunoglu, Stephan Mintz, and Arnold Perlmutter
Plenum Press, 1996

Technology for the Global Economic, Environmental Survival and Prosperity
Edited by: Behram N. Kursunoglu, Stephan Mintz, and Arnold Perlmutter
Plenum Press, 1997

25th Coral Gables Conference on High Energy Physics and Cosmology
Edited by: Behram N. Kursunoglu, Stephan Mintz, and Arnold Perlmutter
Plenum Press, 1997

Environment and Nuclear Energy
Edited by: Behram N. Kursunoglu, Stephan Mintz, and Arnold Perlmutter
Plenum Press, 1998

CONTRIBUTING CO-SPONSORS OF THE GLOBAL FOUNDATION CONFERENCES

Electric Power Research Institute, Palo Alto, California
Gas Research Institute, Washington, DC
General Electric Company, San Jose, California
Northrop Grumman Aerospace Company, Bethpage, New York
Martin Marietta Astronautics Group, Denver, Colorado
Black and Veatch Company, Kansas City, Missouri
Bechtel Power Corporation, Gaithersburg, Maryland
ABB Combustion Engineering, Windsor, Connecticut
BellSouth Corporation, Atlanta, Georgia
National Science Foundation
United States Department of Energy

CONFERENCE PROGRAM

SUNDAY, October 26, 1997

1:00 PM - 9:00 PM REGISTRATION - ***Hamilton Room***

MONDAY, October 27, 1997

8:30 AM **SESSION I:** **BASIC REQUIREMENTS FOR AN ACCEPTABLE GLOBAL NUCLEAR ENERGY FUTURE**
A. The Role of Nuclear Power

Opening Remarks: **BEHRAM N. KURSUNOGLU,** Global Foundation, Inc., Coral Gables, FL

Moderator: **RICHARD KENNEDY,** Washington, DC

Keynote Dissertator: **CHAUNCEY STARR,** EPRI
"Energy Alternatives And Global Warming In The 21st Century"

Keynote Address: **HANS BETHE,** Cornell University
"The Necessity Of Nuclear Power"

Annotators: **MELVIN S. COOPS,** Lawrence Livermore National Laboratory
CHARLES YULISH, US Enrichment Corporation, Maryland

Session Organizer: **BEHRAM N. KURSUNOGLU**

10:30 AM Coffee Break

10:45 AM **SESSION I:** **BASIC REQUIREMENTS FOR AN ACCEPTABLE GLOBAL NUCLEAR ENERGY FUTURE - CONTINUED**
B. Nuclear Power Reactors, Safety

Moderator: **BERTRAM WOLFE,** Former Vice-President, GE Nuclear Division, California

Dissertators: **GLENN T. SEABORG,** Lawrence Berkeley National Lab
"New Directions in the Development of Nuclear Power"
EDWARD TELLER, Lawrence Livermore National Laboratory, University of California
"A Positive Approach to Nuclear Reactors"
GARY HOLAHAN, US Nuclear Regulatory Commission, Washington DC
"Reactor Systems Safety and Analysis"

Annotators: **JOHN IRELAND,** Los Alamos National Laboratory
JOSE G. MARTIN, University of Texas at Brownsville
WILLIAM SUTCLIFFE, Lawrence Livermore National Laboratory

Session Organizer: **WILLIAM JACOBI,** Monroeville, Pennsylvania

12:30 PM Lunch Break

MONDAY, October 27, 1997

1:30 PM	**SESSION II:**	**CHALLENGES OF NUCLEAR AND FOSSIL FUEL WASTES AND THE ENVIRONMENT** **A. Waste Generation from Fossil Fuel and Nuclear Power Use**
	Moderator:	**SHELBY T. BREWER,** S. Brewer Enterprises, Inc. , Connecticut
	Dissertators:	**STEVE FETTER,** University of Maryland "Energy and Global Warming" **DIANA MACARTHUR,** President's Committee of Advisors on Science and Technology (PCAST) "Highlights of the Energy R & D Report" **CURT MILEIKOWSKY,** Switzerland "Improving The Environment Friendliness Of Nuclear Power And Of Fossil Power"
	Annotators:	**MARCELO ALONSO** , Florida Institute of Technology **TOSHIHIKO OKUNO,** Tohoku Electric Power, New York
	Session Organizer:	**JOSE G. MARTIN,** University of Texas at Brownsville
3:15 PM	Coffee Break	
3:30 PM	**SESSION II:**	**CHALLENGES Of NUCLEAR AND FOSSIL FUEL WASTES AND THE ENVIRONMENT - CONTINUED** **B. Fossil Fuel and Nuclear Power Wastes Corresponding to Energy Generation Spanning One Hundred Years**
	Moderator:	**EDWARD ARTHUR,** Los Alamos National Laboratory
	Dissertators:	**EDWARD ARTHUR** "Character and Inventories of Nuclear Materials and Wastes for Possible Future Energy Scenarios" **RICHARD MYERS,** Nuclear Energy Institute, Washington, DC "Prospects for Nuclear Power in a Competitive Environment" **WILLIAM MAGWOOD,** U.S. Department of Energy, Washington DC "Status of DOE's Office of Nuclear Energy's R&D Activities"
	Annotators:	**MAKOTO KAJIKAWA,** The Federation of Electric Power Companies of Japan, Washington, DC **RICH VON KONYNENBURG,** Lawrence Livermore National Laboratory
	Session Organizer:	**ANTHONY FAVALE,** Northrop Grumman Corporation
6:00 PM	Conference Adjourns For The Day	
6:00-7:30	***Welcoming Cocktails Party And Art Exhibit, Pre-Function Area***	

TUESDAY, October 28, 1997

8:30 AM	**SESSION III:**	**GLOBAL ENERGY AND ENVIRONMENTAL CHALLENGES ARISING FROM THE DEMANDS OF INDUSTRIALIZATION OF THE REGIONS WITH LARGE POPULATIONS** **A. How Would A Global Use Of Nuclear Energy Impact Its Competitiveness Relative To Other Energy Alternatives**

Moderator: **PIERRE ZALESKI**, University of Paris at Dauphine **Dissertators:** **BERTRAND VIEILLARD-BARON**, Fromatome, Paris
"Some ideas about the competitiveness of nuclear energy in rapidly developing countries"
ROBERT A. KRAKOWSKI, Los Alamos National Laboratory
"Nuclear's Role in Mitigating Greenhouse Gas Emissions"
ROBERT N. SCHOCK, J.A. HASSBERGER, T.H. ISAACS, M.S. STRAUCH, Lawrence Livermore National Laboratory
"Prospects of and Requirements for Nuclear Power as a Contributor Toward Manageing Greenhouse Gases"

Annotators: **ARNOLD BAKER**, Sandia National Laboratory, New Mexico

YUICHIRO MATSUO, The Federation of Electric Power Companies of Japan, Washington, DC

Session Organizer **MARCELO ALONSO**, Melbourne Beach, Florida

10:30 AM Coffee Break

10:45 AM **SESSION III:** **GLOBAL ENERGY CHALLENGES ARISING FROM THE DEMANDS OF INDUSTRIALIZATION OF THE REGIONS WITH LARGE POPULATIONS - CONTINUED**
B. The Economics And Politics Of Nuclear Energy

Moderator: **MYRON KRATZER**, Washington, DC

Dissertators: **CLINTON BASTIN**, U.S. Department of Energy
"Total Quality Corporate Management and Total Quality Government Policies - Essential for Nuclear Technology and Sustainable Energy for a Clean Environment"
WOLFGANG KREÖGER, Paul Scherrer Institut, Switzerland
"Sustainable Development of Energy Supply"
SHELBY T. BREWER
"Nuclear Power as a Business: Financing the Nuclear Renaissance"

Annotators: **MARIE AGNES NDIAYE**, COGEMA
DAVID BROWN, The Stanton Group, Boston, Massachusetts

Session Organizer: **EDWARD ARTHUR**

12:30 PM Lunch Break (*12:00 - Lunch Meeting of the Global Foundation Board of Trustees and Advisors - Farragut Room*)

1:30 PM **SESSION IV:** **NUCLEAR ENERGY: ECONOMIC, POLITICAL, AND ENVIRONMENTAL CONSIDERATIONS**
A. Nuclear's Impact on The Environment

TUESDAY, October 28, 1997

Moderator: **J. ARTHUR DE MONTALEMBERT**, COGEMA, France

Dissertators: **WILLIAM MARTIN**, Washington Policy and Analysis, Inc.
"The Contribution of Nuclear Energy to Global Economic and Environmental Needs"
JUAN EIBENSCHUTZ, Luz y Fuerza del Centro, Mexico
"International Nuclear Utility"
STEFAN HIRSCHBERG, Paul Scherrer Institut, Switzerland
"Integral Assessment of Energy Systems"
JEAN-PIERRE ROUGEAU, COGEMA and the Uranium Institute
"The Importance of Nuclear Energy to Sustainable Development"

Annotators: **ALAN PASTERNAK,** Lawrence Livermore National Laboratory, California

Session Organizer: **PIERRE ZALESKI,** University of Paris at Dauphine

3:00 PM Coffee Break

3:15 PM **SESSION IV:** **NUCLEAR ENERGY: ECONOMIC, POLITICAL, AND ENVIRONMENTAL CONSIDERATIONS - CONTINUES**
B. Evolution Of The Environment

Moderator: **PAUL PORTNEY,** Resources for The Future, Washington, DC

Dissertators: **PAUL PORTNEY**
"The Social Cost of Electricity Generation"
MICHAEL T. WOO, US Enrichment Corporation, Maryland
"Political Considerations"

Annotators: **EARLE MARIE HANSON,** Los Alamos National Laboratory, New Mexico
L. MANNING MUNTZING, Morgan, Lewis & Bockius LLP, Washington, DC

Session Organizer:

5:00 PM Conference Adjourns For The Day

7:00 PM *CONFERENCE BANQUET - HAMILTON BALLROOM*

After Dinner Speaker: **HANS BLIX,** Director General of the International Atomic Energy Agency, Vienna Austria
"Environment, Nuclear Energy, And Public Perception"

WEDNESDAY, October 29, 1997

8:30 AM **SESSION V:** **ENERGY FUTURE OF THE WORLD**
A. Can The World Sustain Its Economic And Environmental Needs Without Nuclear Energy?

Moderator: **SCOTT CAMPBELL,** Washington Policy and Analysis, Inc.

Panel Members: **KURT MILEIKOWSKY,** Stockholm, Sweden; **PAUL PORTNEY; DAVID ROSSIN,** CISAC, Los Altos Hills, California; **CHAUNCEY STARR,** EPRI; **JAMES TAPE; RICHARD WILSON,** Harvard University;

Annotators: **MATHEW MCKINZIE,** Natural Resources Defense Council, Washington DC
DIANNE D'ARRIGO, Nuclear Information and Resource Service, Washington DC

WEDNESDAY, October 29, 1997

DAVID CULP, Plutonium Challenge, Washington DC
(The annotators in this session are expected to participate in the discussions representing the viewpoints of their institutions.)

Session Organizer: **BERTRAM WOLFE**

9:45 AM Coffee Break

10:00AM **SESSION V:** **ENERGY FUTURE OF THE WORLD**

B. Can The Use Of Nuclear Energy Be Globalized With Built-In Security And Safety Measures?

Moderator: **RICHARD WILSON,** Harvard University

Panel Members: **HANS BETHE,** Cornell University; **MYRON KRATZER; PAUL LEVINTHAL;** Nuclear Control Institute, Washington DC; **JEAN-PIERRE ROUGEAU; EDWARD TELLER; RICHARD WAGNER; BERTRAM WOLFE; PIERRE ZALESKI**

Annotators: **SCOTT DENMAN,** Safe Energy Communication Council, Washington DC
(The annotators in this session are expected to participate in the discussions representing the viewpoints of their institutions.)

Session Organizer: **JOSEPH D. LEHMAN,** Martin Marietta Astronautics Group, Denver

11:30 AM **PRESS CONFERENCE**

12:30 PM 1997 CONFERENCE ADJOURNS

CONTENTS

A BRIEF REPORT ON GLOBAL FOUNDATION'S 20TH ENERGY CONFERENCE

Behram N. Kursunoglu
Global Foundation
P.O. Box 249055
Coral Gables, Florida 33124

The 1997 conference took up the most timely topic, "Environment and Nuclear Energy," and was again held in Washington, D.C. It emphasized the impact on environment of the use of energy. Could nuclear energy be a major contributor to an energy source mix that can help to alleviate greenhouse gas emissions and global warming problems? To what extent is it true that the use of nuclear energy in all regions of the world could help to solve pollution problems such as acid rain arising from heavy use of fossil fuels, particularly coal. Do the issues of nuclear waste and a possible nuclear weapons proliferation present scientific, technological, and political challenges? If all these problems inhibit the global use of nuclear energy then, the world is facing an important global issue requiring global solution.

The Conference examined the impact of nuclear energy on regional and global environmental issues under a variety of scenarios that include (1) competition in deregulated energy environments; (2) constraints levied upon use of fossil energy; and (3) possible expansion of nuclear into energy sectors beyond generation of electricity, e.g. process heat, fuels production, etc.

Closely coupled with the examination of environment and nuclear energy was the assessment of the overall role of nuclear energy in meeting future energy needs arising from growing world populations and economic development. Similarly, the role of nuclear power in meeting national and regional energy security objectives was addressed.

In many respects, this conference has had some historic occasions, more than the previous conferences. In particular, seeing Hans Bethe and Edward Teller, who rarely agree on anything, sitting side by side in the front row was viewed by the participants as a remarkable occurrence. Their presence in this conference greatly increased its visibility.

I prepared a white paper and circulated it among members of the Committee, which included Chauncey Starr, Bertram Wolfe, Edward Arthur, Richard Kennedy, Anthony Favale, Richard Wilson, Pierre Zaleski, and Glenn Seaborg. Their viewpoints and comments were taken into consideration while revising the white paper. I have received a letter from Edward Teller delivered by Federal Express indicating that he would not sign it. At first I thought that this was unfortunate, since his name would add to the credentials of this white paper. I decided to revise it and, during the meeting in Washington, I asked Hans Bethe if he would consider reading and making suggestions on the paper, which he accepted. The

Environment and Nuclear Energy
Edited by Kursunoglu *et al.*, Plenum Press, New York, 1998

resulting white paper was shorter than the original one and was more succinct. I sent it to Edward Teller for his comments. Of course, I did not tell him that Hans Bethe had anything to do with it. This was the strategy we decided (with Bethe) to get Teller's agreement. He, the next day, decided to sign up, even though he wanted some changes but I persuaded him not to change anything. I have also asked Glenn Seaborg to comment and accept to be one of the signatories. That worked out very well and we had, essentially, an important document. The remaining signatories have accepted, after making some suggestions, to sign up also. The Nuclear Energy Institute in Washington volunteered to help us to deliver 500 copies to all members of the United States Congress.

The white paper was released at a press conference on October 29, 1997, at 11:00 AM, in the Crown Plaza Hotel. There were approximately 20 participants from the press. We were competing for press attention with the visit by Chinese President Jiang Zemin and members of the Chinese delegation, who were negotiating that same day with President Clinton to buy Nuclear Reactors from the United States. The entire press conference was organized by Laurie Cunnington, a very close friend of my family. This was the fourth time she has organized a press conference for our fora. The press conference was covered extensively in the October 30th and November 6th issues of Nucleonics Week. There were articles in other papers and a report on National Public Radio (NPR). Because of its importance, I would like to include the white paper at this point.

WHITE PAPER

To be released during the **Global Foundation** Press Conference
October 29, 1997 at 11:30 am in the Crowne Plaza Hotel ,Washington D.C.

On the occasion of the **Global Foundation's** 20th Energy Conference, "Environment and Nuclear Energy", held in Washington, DC, from October 27-29, 1997, the undersigned have considered global energy needs for the future and, also, world development to meet this demand in an environmentally acceptable way.

1. Energy needs will increase throughout the world, particularly in developing countries due to the combination of growing populations and industrialization.
2. In these countries a major energy source will, of necessity, be fossil and organic fuels which will increase emissions of greenhouse gasses. This will compound already significant worldwide environmental problems.
3. Energy conservation in developed countries cannot adequately offset the growth in energy use by developing countries . The developed countries must, therefore, put increased emphasis on non-fossil energy technologies.
4. Although technological innovation may eventually provide non-polluting alternatives, at present only nuclear power is a cost effective non-fossil source of electric power.
5. It is therefore vital that the United States in particular, and all developed countries, emphasize nuclear power in meeting electric power needs, and to the extent possible substitute uranium for fossil fuel. It is equally critical that, as aging nuclear facilities are taken out of commission, replacement power generation be nuclear and not fossil fuel.

6. While we recognize the major concern attendant on widespread use of nuclear power: in particular reactor malfunction, we note that no reactor accident that harmed any member of the public has occurred in any facility meeting international safety standards (Chernobyl did not meet the standards). Fossil fuel pollution from power plants is estimated to cause 40,000 to 70,000 deaths per year in the United States alone.
7. The issue of nuclear weapons proliferation can be met, we believe, by strengthening the International Atomic Energy Agency both scientifically and by providing it with means of enforcement.
8. Technology exists to dispose of nuclear waste safely.
9. We hope that the Kyoto meeting will call upon all countries to cooperate in deployment of nuclear power as the available means of responsibly meeting the world's energy needs. The issues are of a global extent seeking intelligent international cooperation.

Signatories for the Global Foundation White Paper
October 29, 1997
Crowne Plaza Hotel, Washington, DC

Edward Arthur
Senior Science Adviser, Nuclear Materials and Stockpile Management, Los Alamos National Laboratory

Richard Kennedy
Formerly Member ,US Nuclear Regulatory Commission, Ambassador At Large to IAEA Vienna

William Martin
Formerly Deputy Secretary, US Department of Energy, Chairman Washington Policy & Analysis

Chauncey Starr
President Emeritus and Founder of the Electric Power Research Institute

Edward Teller
Formerly Associate Director of Los Alamos National Laboratory, and Director Emeritus Lawrence Livermore National Laboratory University of California

Bertram Wolfe
Formerly Vice President and General Manager General Electric Nuclear Energy

Hans A. Bethe
Nobel Laureate in Physics
Cornell University

Anthony J. Favale
Director, Advanced Energy Systems
Northrop Grumman Corporation

Behram N. Kursunoglu
Theoretical Physicist
Chairman of the Board, Global Foundation, Inc., Professor and Director Emeritus
University of Miami

Glenn T. Seaborg
Nobel Laureate in Chemistry, University of California at Berkeley, Formerly Chairman, U.S. Atomic Energy Commission

Richard Wilson
Mallinckrodt Professor of Physics
Harvard University

Pierre Zaleski
Center for Geopolitics of Energy and Natural Resources, Universite Paris Dauphine
Formerly Nuclear Energy Attaché of the French, Embassy of Washington, DC

PRESENTATIONS ON THE INTERNATIONAL CONFERENCE ON ENVIRONMENT AND NUCLEAR ENERGY

Robert A. Krakowski

Systems Engineering & Integration Group
Los Alamos Scientific Laboratory
Los Alamos, New Mexico 87545, and

José G. Martín

College of Mathematics, Science and Technology
University of Texas at Brownsville
Brownsville, Texas 78520

INTRODUCTION

There are complex global connections between economics, energy, and the environment. Meeting the rising expectations of a growing world population depends on economic growth, but if that growth is fuelled by fossil fuels, there are dire implications in terms of the increase in carbon emissions, the changes in the acidity of the rain, and the living conditions in our planet.

The Global Foundation, Inc., a non-profit organization that examines the global implications of those connections, organized this Conference on the Environment and Nuclear Energy to focus on the role that the nuclear energy option may play in affecting those linkages. In addition, the Conference was to examine the issues of nuclear waste and a possible weapons proliferation.

That examination must take into account regulatory climates, economic environments, external constraints, public perception and acceptability, security issues, and market and technological expansion into energy sectors beyond electricity generation. The Conference was to provide a forum for the difficult discussion of these interrelated issues. It was organized in the form of five topical sessions that will be summarized below. The high caliber and intellectual rigor of the presentations will be obvious from the contributions listed here.

Before proceeding to those summaries, it is proper to give some perspective. Some of the objectives of the conference were only partly met: the examination of schemes for non-electric nuclear energy utilization and the imposition of constraints on fossil fuel burning are two examples. Also, stretching the imagination a bit, there are advanced concepts that have the potential of changing the linkages between economics, energy use,

Environment and Nuclear Energy
Edited by Kursunoglu *et al.*, Plenum Press, New York, 1998

and environmental impact in revolutionary ways. Is it feasible to achieve a nuclear energy system with no long-term radioactive waste stream? Are there truly sustainable nuclear options based on the thorium cycle, accelerator-driven systems, and/or fusion-fission hybrids? Is there a role for partial separations and "dry" processing to help close a proliferation-resistant fuel cycle? Is there a role for advanced fission reactors? Is it feasible or important to quantify proliferation risk and potential from the civilian fuel cycle? In an internationalized and sustainable fuel cycle, is there a role for permanent repositories, concepts based on International Monitored Retrievable Surface Storage and "stowaway" (rather than "throwaway") spent-fuel options? Anyone expecting a future-looking discussion of these potentially revolutionary concepts might have felt disappointed.

There is another caveat. The presenters were recognized experts in the field of nuclear energy. They understand the nuclear energy solutions. Justifiably, their advocacy positions rested firmly on the side of nuclear energy. The presentations were rich in projections of historical perspectives and the status of nuclear energy from the perspectives of advocates who worked long and hard in this area. Most presenters understood the so-called "for nuclear energy nemeses" – safety, waste, proliferation, economics – and they firmly believed that technical solutions to these challenges existed.

It may be argued that the fact that there were few participants who opposed the nuclear option was to be expected because those who understand the science and the technology cannot reasonably oppose the option. On the other hand, it was also apparent that few presenters would know how to deal with the adverse public reaction that these "four nemeses" has generated. It is this reaction that explains the lack of acceptance by those who hold the keys to the barriers.

These caveats are not intended to overshadow the excellent points made through the conference. There was a tremendous amount of excellent information, valuable insights, and plenty of "food for thought." The summaries make this point emphatically.

THE MEETINGS

SESSION I

BASIC REQUIREMENTS FOR AN ACCEPTABLE GLOBAL NUCLEAR ENERGY FUTURE

A. The Role of NuclearPower

Moderator and Organizer: Dr. Behram N. Kursunoglu, Global Foundation, Inc P.O. Box 249055, Coral Gables, Florida 33124-9055.
Annotators: Dr. Melvin S. Coops, Lawrence Livermore National Laboratory, P. O. Box 808, Livermore, California.
Mr. Charles Yulish, US Enrichment Corporation, 6903 Rockledge Drive, Bethesda, Maryland 20817.
Keynote Dissertator: Dr. Chauncy Starr, Electric Power Research Institute, 3412 Hillview Avenue, Palo Alto, California 94304-1395.

"Energy Alternatives and Global Warming in the 21st Century"

This wide-ranging presentation first noted that the world of "real politics" has shifted to a view of environmental effects in which the atmosphere is a "global commons" in which control can be implemented through economic (e.g., policy, taxes, etc.) and technological means. Despite this view, Dr. Starr argued that no group has the foresight to plan the global future on the basis of "predictions" of econometric models.

Replacement of fossil fuels by nuclear energy is totally impracticable: such as substitution would destroy the economic base on which the expansion of nuclear energy would depend. A balanced and long-term plan is needed to effect the needed changes. At present, politics feed public fears through the "mantra" of global warming: "populism is overcoming sweet reason".

The basis of energy demand is driven by the developing countries, and understanding the E^3 trilemma (energy-economics-environment) is the key to achieving balance and the basis for reasonable discourse. With respect to population (public health, fertility), pollution (degradation, inefficiency) and prosperity (economics and politics), each global region establishes a different relational equilibrium that defines the particular "human state." This evolving balance is largely determined by the competition of capital for improved efficiency and productivity, and capital directed at the supply side of the balance. Capital supply per se is not a problem - capital allocation is.

Nuclear energy is the only non-carbon energy source that can meet future demand; other renewable energy sources cannot. Biomass conversion makes no sense outside of a region with a radius of 30-40 km around the power plant; this distance define the economic breakeven point with respect to the worth of the fuel. Hydroelectric resources are limited globally. The cost of solar energy (both photovoltaic and wind) is made high by the intermittance of the source and the need for expensive energy storage.

For nuclear energy to flourish, however, there must be long-term stability and security of the financial markets.

Of the four main concerns and issues affecting nuclear energy utilization (safety, waste, cost, proliferation), proliferation looms largest in the public mind, even though the civil nuclear fuel cycle is far from the first choice of material for nuclear proliferation – by a long shot.

While all non-carbon-based energy sources (renewable and nuclear) will be needed to stem global warming, there is no realistic proposal to stabilize atmospheric inventories of greenhouse gases. Most proposed plans can only delay the increase in CO_2, and humanity must eventually accommodate to whatever change occurs in the environment. More importantly, society must maintain the economic and technological infrastructure needed to meet the problems arising from those changes. Given that oil resources are limited and prices will eventually rise, it is the responsibility of industrialized countries to advance the nuclear energy option for eventual use in developing countries in an era of diminishing resources.

Keynote Address:

Dr. Hans Bethe, Division of Nuclear Studies, Cornell University, 320 Newman Lab., Ithaca, New York 14583

"The Necessity of Nuclear Power"

It may be stated that nuclear energy is "in trouble" in the U.S., because no new nuclear power plants have been built since 1972, old orders have been cancelled, and the issues of construction delays, high costs and uncertainties have not been controlled. Still, the fact remains that there are still 110 plants in the country; that those plants are the source of 21% of all the electricity generated here, and that those plants operate with more than 80% plant availability.

Nuclear energy is not in trouble everywhere. Notably, the French experience is quite different. In France, 80% of the electricity is generated by nuclear plants, the costs are favorable, and electricity is exported to other countries. While there is only one supplier, nuclear plants are standardized and prefabricated, and the designs are gradually evolving and improving.

The East Asian economies are expanding, and this provides opportunities for nuclear energy utilization. Japan, for example, has little fossil fuel resources, and supplies and sources of uranium are widely spread across the globe, ensuring availability at favorable costs.

With respect to the environment, although the greenhouse warming is a fact, the U.S. has not addressed the one energy source that is commercially developed and emits little or no greenhouse gases. Particulates from fossil fuels kill an estimated 40-70,000 persons each year: this is about three times the number of deaths from the Hiroshima and Nagasaki bombings; the increase in cancer deaths from radiation doses of 20-50 rads in those attacks was approximately ~1.5%. There is no proof that low radiation doses may be harmful – and some indication that they may actually be beneficial to human health.

The two major nuclear power plant accidents experienced over the last two decades must be viewed quite differently. In the case of Three-Mile Island, the company lost about two billion dollars, but iodine and cesium were trapped in the reactor cooling water and were not released. From a conservative estimate based on the linear dose-effect hypothesis, the radiation from this accident caused about ½ cancer. The damage from this accident was much less than predicted by the Rasmussen Report. The Chernobyl accident, of course, is a different story. Thirty-one deaths occurred immediately (within four months of the explosion), and 200 persons experienced radiation sickness. The accident caused 550 thyroid cancers (children drinking iodine-contaminated milk), and it is not known how many other types of cancer: the incubation period for leukemia is about ten years. About 400,000 persons had to be evacuated. There is no sense to this tragedy: a reactor such as Chernobyl should never have been: it was actually designed in such a way that the loss of water coolant forced prompt criticality. There is little in common between this horror and the Three Mile Island accident, except for one fact: both accidents were caused by inexperienced crews.

In terms of public perception and acceptance, waste disposal is the main challenge facing nuclear power. Based on this perception, the NIMBY ("Not In My Back Yard") syndrome is preventing a rational resolution to this problem in the U. S. In fact, however, defensive in-depth procedures may ensue that no radioactive is released to the atmosphere o the biosphere. Nature has provided proof of long-term containment: at the site of the Okla uranium reactor, in Gabon, Africa, that went critical over 1800 million years ago, there is hardly any indication of fission-product movement relative to the ^{235}U product of plutonium decay.

Corrected by inflation, the cost of power generated by nuclear plants increased by nuclear power plants over the period 1973-88 increased by ~2 during the period between 1973 and 1988. This increase was driven by delays in construction and the cost of money during construction. In terms of capital costs, average estimates are ~\$2/We; but they can be as high as ~\$5/We.

Positive indicators abound, however:

a) New LWR designs (i.e., AP600) have a simplified design and are "passively safe" - cooling by natural convection is adequate to prevent a core melt, once the core is shut down,
b) Operation and maintenance costs have decreased from three to two mill/kWeh – the same as coal-fired plants); and
c) With capital costs of \$2/We , at 10% interest rate, and with 70% plant availability, the cost of generated nuclear power is about 33 mill/kWeh, which is competitive.

The big question is: what will take the place of existing nuclear plants, when they "retire"? Will they be fossil power plants, or nuclear plants?

Secondly, there is a need for simple, deployable (with respect to local infrastructures), and easily manageable nuclear plants. Will they be available?

The last great question is: Why is nuclear energy not an important part of the ongoing discussions on greenhouse warming and global climate change?

The public remains unaware of the actual and predicted safety record for nuclear plants: the AP600 is projected to have a plant failure once each 800,000 years; nuclear waste transport is actually safer than gasoline transport, etc. Public education is sorely needed, but *"I do not know how to educate the public."* One last comment on the waste: from the perspectives of resources, repository site, and the public perception, only fission products should be buried. All uranium and transuranics should be burned.

B. Nuclear Power Reactors, Safety

Moderator: Dr. Bertran Wolfe, GE Nuclear Division, 15453 Via Vaquero, Monte Sereno, California 02138

Organizer: Dr. William M. Jacobi, 119 Mount Vernon Drive, Monroeville, Penssylvania 156146

Annotators: Dr. John R. Ireland, Los Alamos National Laboratory, Mail Stop F606, P. O. Box 1663, Los Alamos, New Mexico 87545

Dr. José G. Martín, University of Texas at Brownsville, Brownsville, Texas 78520

Dr. William Sutcliffe, Center for Technical Studies, Lawrence Livermore National Laboratory, P. O. Box 808, L-175, Livermore, California 94550

Dissertators: Dr. Glenn Seaborg , Lawrence Berkeley Laboratory, 1 Cyclotron Road, Building 70A/Room 3307 Berkeley, California 94720.

"New Direction in the Development of Nuclear Power"

(Note: This paper was read by Dr. Wolfe from a manuscript submitted by Dr. Seaborg; Dr. Wolfe also made some introductory remarks. He stated that the main cause of the stagnation of nuclear energy in the U. S. is political, and that this type of politics has thrived in an environment where there is an electricity generation surplus. Because of the politics, there are unnecessary construction and licensing delays penalize the economics of nuclear power. Technology is not the issue: in Japan, for example, the institutional climate makes it possible for nuclear plants to be built in less than four years. Nuclear energy in the U. S. will not have a renaissance until there are no more unreasonable court delays and the one-step licensing process is demonstrated to work. Until then, no U. S. utility will commit to a nuclear power plant. In fact, most U.S. utilities with nuclear plants would drop them if they could – to run those plants is too much trouble in the present "over-regulatory" environment.)

The "new directions" addressed in Dr. Seaborg's address were the ABWR and the System 80^{+} systems, that received NRC standardized reactor approval; the AP600, presently under review; and the ALMR and the PRISM reactors. All of these "inherently safe" reactors are characterized by simplified design characteristics.

Considering other alternatives, the paper related some of the problems associated with fusion power: the challenge of how to contain thermonuclear plasmas, the need for very long-term research and development, and the high radioactivity. [The radioactivity issue may be circumvented by the use of colliding beams for the fusion of neutron-free fuel cycles (such as the proton-^{11}B cycle) with direct conversion of charged particles in a compact and safe reaction system without residues. Because of radiation losses, Dr. E. Teller doubts the feasibility of accelerator-beam fusion – apparently it took him a week to convince Fermi that it would not work. Lowell Wood, on the other hand, claimed that high-brightness beams without electrons (e.g., with no radiation) might work, although the concept would not be easily achieved. A concept that Dr. Seaborg seems to favor is the MIGMA type, which has problems of its own.]

High cost and the fear of proliferation were identified as the main obstacles that stand in the way to a nuclear contribution to meet the energy needs of developing countries. It was noted that the world is in an "energy residue" rather than in an "energy resource" crisis. That is, the problem is not one of, say, lack of oil; the problems are the greenhouse gases, the waste heat, the radioactive waste, etc.

Although many options exist; it may take 30-40 years to do develop those ideas. (*At the end of this presentation, Dr. Starr expressed his concern about how to secure the long-term funding needed to do that. Dr. Wolfe share those concerns, but L. Wood, giving the LWR experience as an example, felt that nuclear reactors could be developed in considerably shorter time-frames. Dr. W. Sutcliffe felt that concerns about proliferation from the civil nuclear fuel cycle were misplaced. Finally, Juan Eibenschtuz felt that a way must be found to use nuclear energy, and particularly fusion energy, more directly for energy applications.)*

Dr. Edward Teller, Associate Director Emeritus, Lawrence Livermore Laboratory, University of California, P.O. Box 808, Livermore, California 94550.

"A Positive Approach to Nuclear Reactors"

Two inevitabilities make nuclear energy interesting and worth thinking about: greenhouse warming and the limited resource of fossil fuels. Dr. Teller cited four problems, stressing the importance of educating the public:

1) radioactivity, accidents, and long-lived waste
2) the finiteness of the ^{235}U resource
3) the high cost of nuclear energy today, relative to other sources, and
4) nuclear weapons proliferation.

There is a way to address these four problems: a new reactor concept proposed by John Nuckolls, Lowell Wood, and Edward Teller. Essential features of the proposed thorium breeder include:

a) startup and operation for ~30 years without operator intervention;
b) underground (~200 m) location under earthwuake-proof loose and dry soil c)
c) high-temperature helium cooling
d) automatic regulation to provide energy on demand
e) hard neutron spectrum

Lithium-6 would be hydraulically pushed into the underground, hard-spectrum core when surface electrical-energy demand dwindled, and similarly withdrawn automatically from the core when demand increased. In the core itself, a (slowly) propagating burn would proceed from the center of the long core towards each end.

It is claimed that this configuration would ensure a much lower probability of accidents. Even if the reactor were to be destroyed, there would be no direct human casualties. Waste products would be disposed of in situ, and the after-heat would keep ground water away for as long a time as necessary.

Thorium resources are huge – they are estimated to be ten times as large as those of uranium. The economics are potentially very attractive, because underegraound excavation is likely to represent at most ~10% of the total capital cost.

Dr. Gary Holahan, Director, Systems Safety and Analysis, United States Nuclear Regulatory Commission, Mailstop 18G15 Washington, D.C. 20555.

Future nuclear plants must be safe and environmentally acceptable; economically competitive; and *perceived by the public* to be safe. After stating these requirements, the speaker presented a historical chronology on safety issues in the U.S., as follows:

a) 1950's - small reactors are built; physics studies are made; there are concerns about core life and damage, but little public concern or interest;
b) 1960's – a variety of small commercial reactors are developed; there are few formal safety requirements; reviews are made by the AEC on a case-by-case basis; and there is some consideration of containment buildings;
c) 1970's - large nuclear plant designs emerge; emergency coolant supply systems (ECCS) are required, and the PRA methodology is developed;
d) 1975 - Brown's Ferry fire occurs;
e) 1979 – There is a partial core melt at TMI-2, and severe accidents become a real concern; and
f) 1980's – Safety requirements (170 new requirements!) are added as a reaction to the TMI-2 accident and other operational experiences; and more rigid and quantitative tools are applied to ensure safety.

At the end of this development, new rules emerged (on ECSS, emergency planning, fire protection, electrical equipment qualifications, pressurized thermal shock phenomena, anticipated transients without SCRAM, etc.). There was an enhanced perception for the potential for severe accidents and public harm, and PRA became widely used to search for severe-accident vulnerabilities. On the positive side, the 1980's saw improved utility performance (fewer transients, more reliable safety systems, increased capacity factors, etc.).

Good operating experience continued in the 1990's; at the same time, several plants shut down before their projected end-of-life. Some progress was made on advanced evolutionary LWR designs and on the one-step regulatory approach for ABB/CE's System 80^{+} and GE's ABWR. Westinghouse's AP600 is presently under review by the NRC.

"Risk-informed regulation" integrated PRA with engineering design and helped improve safety and design decision making, increasing NRC staff efficiency and reducing the burdens placed on NRC licensees where appropriate.

The future will see a continuation of risk informed regulation and the approval of new designs shown to meet safety requirements. The future will also bring hoped-for public acceptance, *if:*

a) Current nuclear plants continue to operate safely (even though the TMI-1 accident occurred ~20 years ago, it is still covered and referred by the media); and
b) Credible, open, independent oversight by the NRC of nuclear activities, *as viewed by the public*, continues.

SESSION II

CHALLENGES OF NUCLEAR AND FOSSIL FUEL WASTE AND THE ENVIRONMENT

A. Waste Generation from Fossil Fuel and Nuclear Power Use

Moderator: Dr. Shelby T. Brewer, S. Brewer Enterprises, Inc., Connecticut.

Mr. Robert Repetto, World Resources Institute.

"The Cost of Climate Protection: A Guide for the Perplexed"

The speaker reviewed the 1997 World Resources Institute report of the above title by by himself and D. Austin. This report collects 162 simulations from 16 leading economic models developed and used by academia, national laboratories, and energy industries. The collected results were expressed in terms of the reductions in gross domestic product predicted to accompany a given reduction of CO_2 emissions relative to each study-specific basis case.

The U.S. goal is to reduce CO_2 emission in 2010 to 1990 levels and subsequently holding emissions constant. To be credible, the claims and counter-claims of this and other scenarios must be based on credible models.

It was found that six key assumptions captured 85% of the variance observed in all analyses examined by the study. Specifically, the following variables capture most of the differences in the relationships of gross domestic product versus CO_2:

1) inefficient economic responses;
2) availability of non-carbon backstop fuels;
3) availability of joint international implementation plans ("trade of credits");
4) effectiveness of revenue recycling (reduction of other business taxes);
5) aversion of damage from air pollution; and
6) aversion of climate-change damage.

In spite of the wide range of predicted economic impacts of CO_2 abatement, two areas of policy agreement emerged:

1) the economic impact will be much more favorable if revenue-raising policy instruments are used to control CO_2 emissions, and the revenues are used to reduce other burdensome taxes; and
2) joint international implementation will reduce the overall cost of achieving atmospheric carbon abatement targets.

Ms. Diana MacArthur, President's Committee of Advisors on Science and Technology.

"Highlights of the Energy R&D Report"

Four energy issues were addressed, with references to the findings of the President's Committee of Advisors on Science and Technology (PCAST): a) end-use efficiency; b) fossil fuels; c) renewable energy; and d) nuclear energy. "Cross-cutting" issues such as the management philosophies of DOE were also addressed.

Increased end-use efficiency (in buildings, in transport, through advanced technologies) was found by PCAST to be the most cost-effective way to combat green house emissions. The route by which these efficiencies are to be improved is driven by public service and public good as well as by business goals – specifically, export goals. The largest share of the Research and Development budget was recommended for this area (an increase of ~100%, from 373 M$ to 755 M$).

It was recommended that Research and Development efforts in fossil fuels be held at present levels, but some re-orientation was suggested. Since the largest hurdles rest with renewable energies, a doubling of the renewable energy budget was recommended, to 559 million dollars by 2003.

The emphasis of this presentation was on nuclear power. The PCAST recommended a doubling of the R&D budget for nuclear energy (to 102 million dollars in 2003) to deal with the problems of the stretching out of the horizon for waste disposal; nuclear weapons proliferation; safety; impact of deregulation and privatization of the electricity industries; and the relative decline in electricity use.

A "Nuclear Research Initiative" was defined in terms of a call for new ideas and to change the present mode of "directed research". Areas of emphasis would be:

a) increased conversion efficiencies;
b) development of proliferation-resistant fuel cycles;
c) development of economic low-power reactor concepts for deployment in developing countries; and
d) on-site surface storage of nuclear waste.

Dr. B. Wolfe argued that PCAST had missed the essential point that the problem with nuclear energy was not one of research and development, but with license renewal through the present NRC procedure: no utility has entered this untested procedure because of the uncertainties in time and costs.

According to the presenter, while DOE is to accept nuclear waste by 1998 in Yucca Mountain, a policy for action is needed. Such a policy on nuclear waste is a prerequisite for nuclear energy to move forward.

A "new business opportunity" exists for new methods applied to the back-end of the fuel cycle, and policy changes on the sale of nuclear power and related technologies to China will soon allow U.S. vendors to export along with Canadian and European enterprises..

The 1995 PCAST recommendation for a 329 M$ fusion R&D budget was met by a 233

M$ congressional action, which seriously limits important activities on reduced-activation materials development, ITER, and research on alternative fusion confinement. . The 1997 PCAST recommended the fusion budget to be increased to 281 M$ by 2002, with the 11-B$ ITER device entering a 3-year post-engineering design activity that would focus on site, prototyping, and cost reduction.

DOE operations could benefit from an oversight committee to facilitate communication among industry, university, and national laboratories. All energy programs should undergo strategic reviews similar to those conducted by the Department of Defense. By 2003, approximately one billion dollars should be added to the DOE budget for energy research and development – this still would represent 0.2% of all energy-related R&D conducted in the U.S.

Dr. Richard Myers, Nuclear Energy Institute, Washington, D.C.

"Prospects for Nuclear Power in a Competitive (Deregulated) Environment"

The Nuclear Energy Institute studies issues of nuclear waste, regulation, and economics.

Deregulation of the electric power industries is a major development driving change: it is occurring in states with high electricity costs (e.g., CA, Maine, MA, NV, NH, NJ, KY, OK, PA, etc.); these states are not "going back."

The prospects for federal legislation in this area are dim: there is no congressional consensus. The issue is much more complex than originally thought, the consequences of a "mistake" on people's lives and the economy are great, and the federal action should be limited to "fixing what it can and leaving the rest to the states".

Some states have mandated "unbundling" of the electric utility industry (e.g., separation of generation, transmission, and distribution functions). This unbundling has a profound impact on nuclear power, particularly as it affectds the disposition of sunk costs and costs of plant decommissioning.

During pre-deregulation days there were protected service areas for generation, transmission, and distribution. Now, energy services may compete for customers across all regions, including gas as well as electrical networks. "Wires and pipes" are becoming retail services. As a consequence, generation has become less attractive than ever, and utilities are "selling out" of generation and focusing on the retail end of the business. The business of electricity is becoming a commodity business, and acquiring the characteristics of such a business: - high competitiion, low entry barriers, and volatile prices. Prices become the sole determinant of market success, and the margins are thin: big price swings lead to large losses and large gains.

The outlook for electricity is one of a very fluid and fast-changing market that is difficult to forecast. A surplus of generation is expected until 2005, at which time reserve margins are expected to fall to 10-15%. Approximately 50-75% of new electrical supplies will derive from coal, with heavy use in the Midwest resulting in pollution (SO_x, NO_x) in the U.S. eastern states. An EPA proposal for more stringent air standards can have a big impact on the market, and could significantly enhance the value of existing nuclear power plants.

Generally, competition is good for business, and nuclear power will do well in a deregulated market. Considerations on clean air, global climate change, and the need to retain nuclear energy to offset growth in air emissions from other sectors of the economy augur well for the prospects for nuclear power.

Furthermore, the economics of nuclear power are improving steadily. Specifically, production costs (fuel and operation and maintenance) are competitive with coal-based plants. Capacity factors are up, to 80% (an increase from ~60% in the 1970s). This alone represents the equivalent of 11 GWe of new non-fossil electrical capacity. Generally, companies will be making business decisions that will have an impact on gas and coal, as well as on nuclear power. The shutting down of 1 Gwe-equivalent of nuclear plants receives big press coverage, but neither the public nor the media is interested on whether one GWe of coal-fired plants is uneconomic.

The U. S. has 107 nuclear units operated undere 43 separate licenses. It is forecast that nuclear power plants generating 6-40 TWe will be shut down for "economic reasons". A strong case can be made for consolidation: while multi-unit operations are difficult to beat, single units are most at risk of being shut down. There are unsolved problems associated with such consolidations, however, to wit: a) unfunded decommission costs; b) ease of license transfer; and b) trade restrictions.

Mr. William Magwood, Associate Director Office of Planning and Analysis, U.S. Department of Energy, NE-1 / 5A-143, 100 Independence Avenue, SW, Washington, D.C. 20585.

"Status of DOE's Office of Nuclear Energy's R&D Activities"

The speaker presented a broadly-based report on present and prospective activities in the Office of Nuclear Energy (ONE) of the DOE.

In the near term (1997-98), the ten-year ALWR program is drawing to a close, with the final year of funding for two evolutionary plants receiving NRC design certification (ABWR and the System 80+), and the AP600 final design approval anticipated in 1998.

The DOE proposed nuclear energy security program for 1998 was not funded by Congress, and DOE is now developing new initiatives for 1999. Encouragement may be drawn from Secretary Pena's letter on September 27, 1997 to Senator Myrkowski, recommending to reinforce R&D to assure that nuclear energy continues to contribute to greenhouse mitigation. The PCAST report is also positive, since it helps government send a message to industry that nuclear energy is important.

The future of ONE lies in developing economic, small, proliferation resistant nuclear plants for deployment in developing countries; and assuring that license renewals move ahead on both technical and political grounds. If license renewals do not happen, many more nuclear plants will be prematurely shutdown. Presently, while the U.S. spends much on R&D on energy, little is devoted to nuclear energy. The huge decrease in nuclear engineering degrees granted by our institutions (down by ~50% in the last few years) is a serious infrastructural concern.

Specific R&D plans for nuclear power include: a)the Nuclear Energy Research Institute:b) the development of high burn-up fuels; c) technologies needed to use existing reactors; d) assistance to university nuclear engineering programs; e) emphasis on proliferation resistance of all nuclear technologies. This year, ONE is making an early case for R&D in nuclear energy to Congress; in the past, Congress had made up its collective mind early with respect to needed government-sponsored R&D.
(Dr. Starr noted that, in planning for the future, two "bombshells" must be recognized: a) greenhouse warming, which will go on for many years; and the shift to private competitive nuclear industries and the possible purchasing by independent operators of nuclear plants that are "hitting the licensing barrier".)

Dr. Edward Arthur, Senior Science Advisor Nuclear, Materials and Stockpile Management, Los Alamos National Laboratory, P.O. Box 1663 – MS F628, Los Alamos, NM 87545.

"Character and Inventories of Nuclear Materials and Waste for Possible Future Energy Scenarios"

The long-term concerns of global energy security and the associated inventories and flows of nuclear materials that accompany any role for nuclear power in the future are receiving special emphasis in the Los Alamos Nuclear Vision Project. As technical basis and background, this presentation used the results of scenario modeling performed in support of the Nuclear Vision Project (to the year 2100) and the IAEA Working Group #1 (to the year 2050), both being presented at the June 3-6, 1997, IAEA/NEA "Symposium on Nuclear Fuel Cycle and Reactor Strategies: Adjusting to New Realities". On the basis of high and low variation with respect to a "business-as-usual" base scenario, a range of global spent-fuel inventories was presented. (Refeence was made to a subsequent presentation at this Conference by R. Krakowski.)

The impact of MOX recycling in LWRs (a factor of 2-3 reduction, not unlike that reported in the IAEA/WG#1 study mentioned above) was reported. Finally, a nuclear fuel cycle strategy reported at the October 5-10, 1997, Global'97 International Conference on Future Nuclear Systems was presented, identifying the following attributes were identified:

a) flexibility if dealing with plutonium;
b) less-expensive processing with reduced waste streams;
c) reduced amounts of materials requiring "external" disposition;
d) further reductions in proliferation potential (e.g., radiation barriers and reduced inventories throughout the fuel cycle); and
e) continued utilization and evolution of the LWR experience base for electric power generation.

SESSION III

GLOBAL ENERGY AND ENVIRONMENTAL CHALLENGES ARISING FROM DEMANDS OF INDUSTRIALIZATION OF REGIONS WITH LARGE POPULATIONS

EMELLIO PANARELLO, Editor, Physics Essays, National Research Council, Room 1200, Building M10, Ottawa, KIA OR6, Canada.

"Nuclear Fusion, the World Environment"

The presentation listed the main world-view reasons for pursuing energy generation from nuclear fusion: a) inexhaustible fuel supply; b) greatest energy release per unit mass of fuel; c) tremendous land-use advantage relative to other renewable energy sources; d) no emission of greenhouse gases during operation; and e) greater flexibility to reduce/control generation of radioactive waste through material tailoring/choices.

It also claimed that fusion energy is a near-term reality, and that a proof-of-principle experiment is within the realm of possibility within 3-5 years at a cost of 5-20 M$. This prognosis compares to ITER (International Tokamak Experimental Reactor), which has its demonstration decades away at a cost of ~11 B$ (and growing).

Confidence in the new approach is such that the Canadian government is promoting 50% of this R&D activity under Canadian-controlled private firms. It was argued that the path of minimum risk in achieving viable fusion energy is: a) to move away from the present (tokamak) direction; b) to encourage alternative thinking; and c) to take advantage of Canadian hospitality for research and development.
(Note: The presentation refers to a concept based on the formation of a magnetically insulated, inertially imploded fusion target. However, the concept was not described in detail at the Conference A variant of this approach is also under study at Los Alamos.)

Dr. Bertrand Viellan-Baron, FRAMATOME-Tour Framatome, 92084 Paris-la-Defense, France.

"Some Ideas About the Competitiveness of Nuclear Energy in Rapidly Developing Countries"

In the space of 25 years, nuclear energy has assumed a significant role in providing electrical energy in OECD and FSU countries. The question was raised as to whether NE will play a similar role in developing countries, which presently use what they can for energy (mainly fossil fuels and hydroelectric). It was argued that energy growth is strong and all resources must be used to meet demand, including nuclear power, MOX recycle, and, eventually, fast breeder reactors.

The issue of renewable energies is one of "old versus new" (hydroelectric is "old") However, potential resources are limited to ~7000 TWeh, compared to 2250 TWeh presently being generated annually. The problem of local pollution and greenhouse gas emissions will force a turning to cleaner forms of energy. (How much do carbon fuels contribute to greenhouse gases? On a gram/kWeh basis, coal/oil/gas = 960-1170 / 690-905 / 400-770) Nuclear energy is the major way to reduce those emissions; over the period 1980-1990, the per capita emission of carbon in France has decreased from 8.5 to 6.0 tonneC/capita, whereas this ratio has increased in most industrialized countries.

Even when natural gas prices are low, as they are now, nuclear power is competitive with power from fossil plants in France (37-38 mill/kWeh versus 36 mill/kWeh.) This may be an important lesson for developing countries where demand is growing fast. Key factors in reducing nuclear costs are:

a) reduced amortized investment (both construction costs and time);
b) extended life;
c) increased availability;
d) improved fuel performance;
e) augmented thermal conversion efficiency;
f) improved maintenance effectiveness; and
g) standardization based on a proven model.

An extensive and thorough database was used to describe the status and the potential of nuclear power in developing countries from the viewpoint of need, implementation, and economic competitiveness. Key trade-offs related to economics, technology, ecology, independence, and public acceptance were noted. When trying to meet energy demands in developing countries with fast-growing energy demands, nuclear power presents many advantages: diversification of sources, stability of cost, low sensitivity to international economic situations, good long-term competitiveness, decreased atmospheric pollution, use of tested models, and incentives for increased technological sophistication.

Dr. Robert Krakowski, Systems Engineering & Integration Group, Los Alamos National Laboratory, MS F607, TSA-3, Los Alamos, New Mexico 87545.

"The Role of Nuclear Energy in Mitigating Greenhouse Warming"

An important part of the nuclear-materials/energy-security focus of the Los Alamos Nuclear Vision Project is the long-term (year 2100), multi-regional (13 regions) modeling of nuclear energy and materials in a context that embeds nuclear market penetration into a consistent long-term picture of the global energy economy.

The modifications to an existing and extensively-used global E^3 model to include a high-fidelity nuclear energy model was described. Also described was the impact of upper-hierarchal global scenario drivers or attributes [population growth, economic productivity growth, non-price efficiency improvements, taxes, nuclear capital and O&M (including fuel cycle annual charges) costs, uranium resource constraints, etc.]

Parametric studies focused on the role nuclear power might play in mitigating greenhouse warming, wherein this role is forced through one of two scenario attributes:

a) carbon taxes to effect a largely supply-side force (e.g., fossil fuels are made more expensive in accordance to carbon content, and the market share for nuclear power correspondingly increases); or
b) demand-side forces that cause changes in the rate at which non-price-induced increases in efficiencies with which secondary energy is used to provide energy services.

The presentation described the impact of both demand-side and supply-side drivers on nuclear demand, associated nuclear material (form- and inventory-dependent) proliferation risk, regional economics and greenhouse warming (change in greenhouse gas emissions rates, atmospheric accumulation, and global temperature rises.

A symbiosis between demand-side and supply-side policy tools reveals itself from these studies: high nuclear energy demands induced by carbon taxes can be partly met by concomitant increases in " energy intensity" (gross national product per primary energy demand) leading to the same or even a reduced global climate impact.

(While attempts to relate these variables (economic, environmental, energy demand, attendant proliferation risk) were generally applauded, concern was expressed over attempts to quantify proliferation risk and/or global warming with explicit metrics.)

Dr. James Hassberger, Lawrence Livermore National Laboratory.

"Prospects of and Requirements for Nuclear Power as a Contributor Toward Managing Greenhouse Gases"

This "top-level" talk identified key drivers that may determine the role of nuclear power in mitigating global warming, including:

a) demand uncertainty;
b) shifts in energy sectors (particularly on transportation);
c) disparate distributions (regarding tensions);
d) technological readiness (including cost and availability);
e) market dynamics (deregulation, privatization); and
f) environmental resolve (internationalization, joint implementation).

These elements relate to and define the risks of and of not pursuing nuclear power.

Dr. Clinton Bastin, NTEU (DOE), 14355 Long Channel Drive, Germantown, Maryland 20874.

"Total Quality Corporate Management and Total Quality Government Policies - Essential for Nuclear Technology and Sustainable Energy for a Cleaner Environment"

This was a generally philosophical and reflective talk, punctuated with sharp examples of wrong turns taken. The theme is amply set out in the extended title.

The history of nuclear power development over the past decades offers many examples of "quality events":

a) the achievements of the du Pont Company and the early production of weapons plutonium (when reactors were built in a few years, not decades);
b) the development of similar reactors by du Pont for tritium production;
c) the deployment of the nuclear navy (more than 200 reactors rapidly and efficiently brought on line);
d) the success of the international nuclear power industry.

Subsequently, however, there were a number of wrong turns:

a) the reliance of the U.S. reprocessing efforts on overly optimistic productivity and cost estimates;
b) the decision for the direct disposal of spent fuel, leading to the problem of plutonium mines of the future;
c) the loss of public trust in nuclear power because technical people and the DOE people did not use the "Quality Principles".

The presenter argued that most nuclear-related DOE functions should be turned over to private industry, who know how to apply the "Quality Principles."

a) To illustrate this argument, appendix material was distributed dealing with:
b) the IFR fuel cycle;
c) the present nuclear proliferation policies in the U.S. - and the need for change;
d) principles of quality for managing/regulating/developing policies for using nuclear technologies; and
e) a top-20 list of extraordinary leadership characteristics.

The following elements were listed as essential to regain and maintain the publics trust in the use of nuclear technologies: a) commit the best nuclear sites and materials; b) find creative corporations to manage the nuclear enterprises now assigned to DOE; c) restructure the nuclear regulatory process to assure management according to "Quality Principles"; and d) establish a U.S. Nuclear Policy Board to meet quality principles.

Dr. Shelby Brewer, CEO, S. Brewer Enterprises Inc. 29 Blue Ridge Drive, Simsbury, Connecticut 06089.

"Nuclear Power as a Business: Financing the Nuclear Renaissance"

Reviewing the past and future of nuclear power, this presentation listed the reasons for the financial and institutional down turn of the nuclear option from the point of view of a business person. None of those reasons are technological.

According to this view, nuclear energy will come back when the market is allowed to operate and the fiduciary skill of the plant owners improves. New reactor designs are not going to change the public perceptions of nuclear power in the U. S.

Different environments may encourage the development of nuclear power: a monopolistic model (a la France); turn-key operations (as in early U.S. industry development); status quo; vendor equity models; nuclear independent power producers.

However, the paper made a call for a "nuclear Adam Smith" approach, with the government staying out of the business. A "nuclear renaissance" is needed wherein the market leads and supports the move to deregulation. The major competitor for nuclear power is natural gas and under-utilized midwestern coal, and new opportunities to compete in the open market should be created, unfettered by national and state governments, with nuclear power operated as a business that is freed of government "claptrap".

(Dr. Brewer's view would represent a purely business one, without a role for market regulations or the perception of a need to address explicitly the public concerns to which those regulations relate.)

Dr. Wolfgang Kreoger, Paul Scherrer Institut, CH-5232 Villegan PSI, Switzerland.

"Sustainable Development of Energy Supply"

There is a long history to the concept of sustainable development: it generated discussions about the use of European forests in 1713. The terms were defined and the general goals formalized by the Burndland Report in 1987; at the UN Rio de Janeiro Summit in 1992, the resulting declaration showed a shared understanding and acceptance of the goals of sustainable development. The scope was extended by the Alliance of Global sustainability (an alliance that included MIT, Tokyo University, and ETH) in 1995.

The extended scope of NE sustainability encompasses fuel (depletion time), land use, environment, human health, societal aspects (public acceptance, economics, proliferation), non-production of non-degradable waste (amounts and confinement times), security of fuel and technology supply, and failure friendliness (particularly with respect to developing country applications). The hurdles placed before nuclear power are defined with respect to these sustainability goals: a) unresolved waste issues; b) safety issues; c) competition issues; d) limited resources (for once-through LWR's); e) dependence on minimal political, financial, and social stability.

The development of a new fuel cycle is badly needed to accommodate the following needs and goals: a) no net resource consumption; b) proliferation resistance; c) burning of long-lived actinides and fission products; d) reduction of security confinement times in the final repositories.

In addition to designs of nuclear plants that exclude catastrophic accidents, fuel cycles should be optimized at various levels: a) optimized LWRs with high-burnup MOX; b) advanced fuel cycles (100% MOX, use of non-fertile-fuel hybrid systems); c) fast-breders and integral/dry processing; d) CAPRA systems.

SESSION IV

NUCLEAR ENERGY: ECONOMIC, POLITICAL, AND ENVIRONMENTAL CONSIDERATIONS

A. Nuclear's Impact on the Environment

Dr. William Martin, Washington Policy Analysis, Inc., 1025 Thomas Jefferson Street, NW, Suite 411 West, Washington, D.C. 200007.

"Reducing CO_2 Emissions: Toward a Sustainable Energy Mix"

While the reduction of CO_2 emissions represents an important benefit for sustainable development, one cannot lose site of other essential ingredients to the stable and sustainable long-term future: environment; national security; economic stability; and public health.

The "Global Energy Model System," from Washington Policy Analysis, Inc., was used to examine aggregated CO_2 contributions from a range of energy-consumption scenarios out to the year 2050.

The "business as usual" scenario is based on coal and oil; under this scenario, releases of CO_2 in 2050 reached 16.9 GtonneC/yr, , compared to 7.1 GtonneC/yr in 1990. Tripling the consumption of "non-carbon" power and increasing the use of renewable energy reduced the 2050 emission levels to 13 GtonneC/yr. Expanding nuclear power generation to 1000 GWe in 2050 led to 11.5 GtonneC/yr, and a concomitant increase in demand-side management resulted in a 2050 CO_2 emission rate of 10.1 GtonneC/yr, which is still 42.3% above 1990 emission levels.

While achieving a sustainable future will require a broad-minded approach by all nations, these serial scenario results shows the difficulty of stabilizing the concentration of CO_2 in the atmosphere.

Combining all the elements of a sustainable development (environment, national security, economic stability, public health) led to the following recommendations of energy mix:

(a) diversify oil supplies away from Persian Gulf suppliers;
(b) develop clean coal technologies;
(c) ensure integrity of natural gas infrastructure at the international level;
(d) develop advanced nuclear technologies while resolving waste, safety, and proliferation issues;
(e) accelerate renewable energy development (solar, wind, biomass);
(f) increase non-price energy efficiency goals by 25%; and
(g) encourage alternative fuels for transportation at high efficiency.

Dr. Juan Eibenschutz, Subdirector de Distribución, Luz y Fuerza del Centro, Mexico, Melchor Ocampo 171 Piso 8, Col. Tlaxpana, 11379 Mexico, D.F.

"International Nuclear Utility"

The globalization of the world economic sectors, including energy, brings new opportunities and new technologies to the nuclear energy field, and a range of operational arrangements varying from monopolies to holding companies can be envisaged.

Dr. Eibenchutz proposed an organizational concept for nuclear power that deals with key issues, including safety and nuclear proliferation, fashioned after that which exists for present-day oil companies. This drive for this organization is the need to keep the nuclear option alive and to deal with present problems of public acceptance, waste, proliferation, economics, etc.

An International Nuclear Utility (INU) concept was proposed, wherein complete services would be provided by local utilities or after the INU has taken charge of existing facilities. Subsidiaries with local utilities would be formed, based on reactor technologies they possess; alternatively, standard plants and associated site qualifications would be offered. This fuzzily defined entity must be both international and publicly owned, would be controlled by an international board, and would be responsible for safeguarding a huge amount of nuclear material.

This presentation conclude with a call for participants to suggest and clarify the conflicts and operational *modus operandi* for what amounts to be an international independent power producer modeled after present-day international oil companies.

Dr. Stefan Hirschberg, Paul Scherrer Institut, CH-5232 Villigen PSI, Switzerland.

"Integral Assessment of Energy Systems"

This presentation described an extensive and detailed life-cycle analysis of regional (Swiss) energy systems on the basis of tonnes CO_2-equivalent per GWeh. Areas of future improvement for coal (e.g., mining, plant, transport), natural gas, oil, etc. were examined using a methodology that generated many indicators that are evaluated using a graphical decision-analysis tool to score a range of categories/criteria on the basis of a subjective weighting process. Criteria use in this user-based study included: cost, NOx, greenhouse gas emissions, land use, accident risk, social economic factors, supply security, and public acceptability. For the weightings used, the nuclear option had favorable scores, but the public acceptance criteria generally carried the greatest weight .

Dr. Jean-Pierre Rougeau, Senior Vice-President, COGEMA and Vice-Chairman, Uranium Institute, 2, rue Paul Dautier, BP No. 4, 78141, Velizy Villacoublay Cedex, France.

"The Importance of Nuclear Energy to Sustainable Development"

The important role that nuclear power can play in the mitigation of global warming is better appreciated now than it was at the time of the 1992 Rio Conference. According to a 1995 World Energy Council study of 1995, by the year 2020, CO_2 emission relative to 1990 will be up by 40%, energy demand will be up by 100%, and population will be up by 50%. A decrease in energy intensity (primary energy per GNP) from 0.42 toe/$ (42 GJ/toe) to 0.2 toe/$ is required to stabilize CO_2 emissions - not atmospheric concentrations).

Worldwide, the 17% contribution from nuclear power avoids 0.63 GtonneC/yr (~10% of total 1990 emission). In Europe, this corresponds to 700 Mtonne CO_2 removed through the use of nuclear power- and this is the equivalent of removing all cars in Europe from circulation.

An increase in the number of power plants from 120 to 160 units by 2020 is feasible. The management of nuclear waste is feasible, easily detected, and well controlled; the presentation argued that it is technically and economically feasible to isolate this waste from the biosphere, and that related technologies are improving (recycle options, transmutation, compaction).

Nuclear energy is reliable and amenable for concentration in urban use areas. Fuel costs represent a small part of energy costs, and are expected to remain so for many years. Renewable energies have problems of large land-use requirements, costs, intermittency, etc., and internalization of the costs associated with fossil fuel burning will increase energy costs.

The energy security advantages of nuclear power (e.g., uranium reserves are more uniformly distributed than fossil fuels) gives a desirable diversity of fuel supply, avoids potential conflicts. Furthermore, uranium is easily stored for years. Uranium resources are both abundant and easily extended. Nuclear power is sustainable and does not interfere significantly with the carbon life cycle.

Dr. Paul Portney, President, Resource for the Future, 1616 P Street , Northwest, Washington, D.C. 20036.

"The Social Cost of Electricity Generation"

This presenter noted that the present discussions of global warming that exclude the nuclear option are beyond understanding. The following list of external costs associated with electricity generation was revealing, even if these estimates do not include global warming effects; they are based on the Greenfield plant.

Source	*External cost*[(a)] *(mill/kWeh)*
pulverized coal	1.3-2.9
PWR	0.2-0.3
natural gas[(b)]	0.2-0.3
oil	0.2-1.7
biomass	1.7-4.2

[(a)]Krapnick and Burtraw, "Social Cost of Electricity," Resource and Energy Economics, 18, 423-66 (1996).
[(b)]combined cycle gas turbine

The high external cost associated with biomsss is related to the high particulate content associated with burning high-moisture, high-organic-content fuels, and the dominance of health effects in determining external costs. Ultimately, however, the public will decide, no matter what these costs.

Dr. Michael T. Woo, U.S. Enrichment Corporation, 6903 Rockledge Drive, Bethesda, Maryland 20817.

"Political Consideration"

This presentation was given from the perspective of a legislative aid, rather than from that of the speaker's present position. Given the physics, engineering, and economics of an ideal reactor (e.g., the concept previously outlined by Teller), nuclear energy is not competitive on a full-cost basis even though it is a success on an operational basis. The crucial waste problem is not solved.

Under the market paradigm, nuclear energy will not forward if it is not economic in a competitive, non-protected cost environment.. The Energy Policy Act of 1992 gave a market-oriented framework that is designed to allow an economic system to succeed; a paradigm shift has occurred from the earlier "command-and-control" system (e.g., "stem-to-stern" control and regulation of nuclear power) to the market framework, according to the speaker. With one-step licensing, a cost-effective regulatory process (results-oriented rather than detail-oriented) is needed. With the elimination of coverage provided by the Price-Anderson act, if an industry is economic it should also be able to absorb risk, but a micro-managed process may not be able to accept the risk.

New plants will unlikely be built if waste is still to reside at the reactor site. The public acceptance problem may be a "chicken-and-egg" problem; what comes first: a congressional/legislative framework or the need for market forces?

SESSION V

ENERGY FUTURES OF THE WORLD AND THE ROLE OF NUCLEAR ENERGY

Panel to address the question:
"Can the World Sustain Its Economic Growth and Environmental Needs without Nuclear Energy?"

Panelists: Mr. A. David Rossin, President CISAC c/o Rossin and Associates, 24129 Hillview Drive, Los Altos, California 94024.
Dr. Chauncey Starr, , Electric Power Research Institute, 3412 Hillview Avenue, Palo Alto, California 94304-1395.
Dr. James Tape, Program Manager –NonProliferation and Arms Control, Los Alamos National Laboratory, P.O. Box 1663, MS-E550, Los Alamos, New Mexico 87545.
Dr. Richard Wilson, Harvard University High Energy Physics Lab, Cambridge, Massachusetts 02138.

Dr. David Rossin:

The problems of nuclear power are political, and the agenda for the political debate is set by the adversaries, not by the technical people. In the past, it was not proper of scientists and engineers to assume an advocacy position; as a consequence, the proverbial "playing field" is tilted in favor of nuclear energy opponents.

Dr. Chauncey Starr: Electric Power Research Institute, 3412 Hillview Avenue, Palo Alto, California 94304-1395.

Since most of the world for at least the next century can do without nuclear power, the answer to the question before the panel is "yes". While energy in general, and electricity specifically, is essential for economic growth, nuclear power is not needed to provided this energy. The fossil fuel supply is adequate for the next 100 years, but when the cost of fossil fuels increase, economic growth will decrease. Environmentally, nuclear energy is benign, and proliferation is not a problem from the civil fuel cycle. However, proliferation remains a great public issue.

Solar, wind, and biomass have real restrictions: solar and wind are intermittent; biomass sources are limited to within a few miles of the power station, else the cost of transport exceeds the value of the fuel. The remaining options are fossil plants and nuclear plants. The impact of greenhouse gases can be limited, but not eliminated; without nuclear power, major problems arise with respect to economic growth and environmental impacts.

Dr. James Tape:

In addition to economic and environmental issues, the important issue of national security must be considered. We must: a) minimize environmental disputes; b) reduce the gap between the industrialized and the developing countries; c) avoid proliferation; d) keep pace with population growth.

In terms of national security, no benefit can be identified in a world where nuclear power is not an option.. Specifically, nuclear power is needed to deal with ex-weapons plutonium; safeguarded nuclear plants are not a proliferation problem; nuclear power aids in sustaining economic growth and, thereby, reducing population growth. Waste emerges from all energy generation options and this in principle is not a matter of choice. No reason can be identified for choosing a world where nuclear option is not available..

Dr. Richard Wilson:

Nuclear energy emits no pollution and no greenhouse gases, and it is sustainable beyond the LWRs. However, it has to be cheap to have impact. When expressed on a $/man-rem basis, the cost of regulation of the nuclear industry indicates a regulatory system gone wrong. Regulations that decrease core-melt probabilities from 1/10,000 to 1/100,000 and the shutting down of reactors for no good reason are Draconian actions. A central question is: how do we prevent the NRC from over regulating? Perhaps the NRC should be sued.

In reply to the comments made by this panel, **Dr. David Culp** (Plutonium Challenge, an arms-control lobbyist group) made the following points:

- The nuclear industry has in fact been setting the congressional agenda, based on the bills being put forth;
- Per unit of energy generated, nuclear energy is the energy source that is most heavily subsidized by the government;
- Indonesia still pursues the nuclear power option, in spite of good fossil fuels resources; the military wants this nuclear capability for reasons other than electricity generation.
- In spite of "over regulation" by the NRC, the NRC Commissioners are those originally supported by the NE industry.

Generally, Dr. Culp felt that a serious disconnect exists between the nuclear world and the rest of the world.

Mr. Mathew McKinsie, of the Natural Resources Defense Council, also responded to this panel and felt that safety, waste, and proliferation remain as main issues for LWR's. Once these issues are resolved, the economics of nuclear power should decide its fate. A concern with the NRC was indicated, as well as the potential for negative feedback related to reduced monies available for safety if nuclear power becomes less economic. Any forced rejuvenation of the nuclear option to combat global warming is not warranted since nuclear power has its own serious waste problem; and capital investments would be taken away from other solutions to the greenhouse warming problem. A concern was indicated, if nuclear power is introduced into developing countries: China has 4-5 reactors on order, but their regulatory staff was claimed to be composed on only seven people. Shouldn't the market value and the motives of the Chinese nuclear program be questioned?

(Interestingly, the response of the panel and other conference attendees to these two annotators focused entirely on the issue of proliferation from the civilian fuel cycle, in spite of the "non-issue" status of this issue throughout the Conference. Generally, it was claimed that the military do not want to deal with nuclear power to gain nuclear materials for their purposes, and no nuclear weapon to date was derived from civil nuclear power plants. Dr. Culp claimed that this was not so for the South African experience, and that a nuclear plant operated under TVA is being considered for tritium production.)

Panel to address the question:

"Can the Use of Nuclear Energy be Globalized with Build-in Security and Safety Measures?"

Panelists: Dr. Hans Bethe, ANS BETHE, Division of Nuclear Studies, Cornell University, 320 Newman Lab, Ithaca, New York 14853

Mr. Myron Kratzer, (retired,) Annapolis, Maryland 21401

Dr. Jean-Pierre Rougeau, Senior Vice-President, COGEMA and Vice-Chairman, Uranium Institute, 2, rue Paul Dautier, BP No. 4, 78141, Velizy Villacoublay Cedex, France.

Dr. Edward Teller, Associate Director Emeritus Lawrence Livermore Laboratory, University of California, P.O. Box 808, Livermore, California 94550

Dr. Richard Wagner, Los Alamos National Laboratory, 409 12th Street Southwest Suite 310, Washington, D.C. 20024.
Dr. Bertram Wolfe, Vice President, General Electric Nuclear Division, 15453 Via Vaquero, Monte Sereno, California 02138.
Dr. Pierre Zaleski, Universite Paris Dauphine, Place du Maréchal de Lattre de Tassigny, 75775 Paris Cedex 16, France.

Dr. Hans Bethe:

This panel question could be met with a positive answer, but the IAEA is not sufficiently funded. Also, although U.S. reactors are safe, the IAEA should inspect our nuclear plants as a good example and to "put teeth" into the inspection mandate.

Mr. Myron Kratzer:

While nuclear plants can be used to support nuclear weapon programs, both from the backend of the fuel cycle and from misuse of enrichment plant, no case is known where nuclear weapons have been developed via the nuclear power route. In other words, nuclear weapons from the civil fuel cycle can happen, but it has not happened; and a strong negative correlation exists.

Instead, a small, dedicated facility has been the preference for weapon plutonium, or enriched uranium production. With respect to the spent-fuel issue, the policy of direct disposal of spent fuel is a misguided policy; such plutonium is not disposed, but becomes a future plutonium mine. Safe storage (stowaway, not throwaway) or reprocessing (when economics dictate) are viable options. A global "plutonium economy" must not be allowed to develop. Instead, plutonium should be concentrated in a few "responsible" countries. The termination of IFR by the first Clinton Administration was a "hair-brained" decision and should be reconsidered as an important means to develop proliferation-resistant fuel cycles; if not, we will have the worst choice – facilitated full separation of plutonium.

Dr. Jean-Pierre Rougeau:

It is surprising that the proliferation issue has dominated these panel "debates", and one should focus on the contribution of nuclear power to "peace on earth" by the development, welfare, and improvement of the living standards of humanity. The use of uranium takes man's energy needs outside the carbon cycle. Uranium is well distributed and thereby its use can avoid conflict.

Waste is not a technical problem, but instead is one of reduced political resolve. Nuclear energy has and will welcome all kinds of safeguard regulations to combat proliferation; Germany has recycled MOX for two decades, and France for one decade, without any problem with proliferation. While a great concern, there are adequate measures available and in place for proliferation control. The same cannot be said of global warming. While man can survive without nuclear power, the "diet is not balanced" and such a balanced energy diet is needed for a better life.

Dr. Edward Teller:

The problem is peace, but the scientist can only lay out alternatives or options that are within the realm of the possible; he cannot solve peace per se. Ultimately, the public must choose the mix between the new and the old ways.

The basic problem is that by 2100 mankind will need ten times the amount of energy it generates today. How much of that will be nuclear? The question is what is the balance, and how does this balance translate in terms of the North-South conflict? Without nuclear power, the "ten-times" scenario is difficult to imagine - energy sources will not survive without enormous cost increases, and the environmental impact will be huge.

The ten-times scenario with nuclear power was addressed in terms of the reactor concept described earlier, with some disappointment being expressed that the idea is not receiving more serious consideration. It was noted that little attention was given at this conference to breeder reactors in general, which will be needed in the next 100 years. One should not criticize the critics, but instead put in the face of criticism a future that is good and acceptable.

Dr. Richard Wagner:

A call was made for the 100-year view, which will have a shifted benefit-to-cost ratio. In this context, North Korea and Iraq are near-term issues. Presently, society is locked between two points of view on the nuclear debate, and progress is not permitted. The nuclear advocates call for a change in the fuel cycle, and the critics oppose and challenge any action that does not encompass the elimination of the nuclear option. The view of Wolfe Haefele seems appropriate, that society needs to buy time to allow uncertainties to be resolved, to develop technologies, to install appropriate development programs, and to provide options for the future.

Dr. Bertram Wolfe:

The status is that nuclear power is not needed in the U.S. and no new capacity in France is expected for a decade. Given, however, that global energy needs will be 3-4 times today's demand by the year 2050 and up by ten times by 2100, the means by which these needs will be met presents an obvious and difficult question.

Supplying ~5000 GWe of nuclear power by 2100 will be difficult, but it can be done. A call has been made for an International NRC to approve new nuclear power plants (from a safety viewpoint) and to expand the IAEA (from a safeguards viewpoint). This internationalization of nuclear power would take the form of centralized manufacturing facilities (e.g., for pressure vessels, steam generators, etc.), limited localized reprocessing, and for the global development of the breeder reactor.

Dr. Pierre Zaleski:

While an international NRC is not needed, a "convention" on nuclear safety already exists. The "take-back" plutonium concept (e.g., plutonium returns back to the country of origin) is good. Consideration should be given to waste disposal facilities in China, the Marshall Islands, and Australia. Generally, alternatives should be compared, the "best" solution identified, and adjustments for new future technologies be made (e.g., cheap, efficient isotope separation will nullify present concerns about plutonium). Finally, societal and institutional issues are crucial for a safe and proliferation-free nuclear option to spread around the world.

An anti-nuclear response to this panel was expressed by **annotator Dianne D'Arrigo** (Nuclear Information and Resource Service) in terms of the following Points, concerns, and uncertainties:

- The waste issue engenders conflict, not "peace", as the NIMBY issue testifies;
- The life of nuclear waste far exceeds the time of institutional control;
- The addition of new facilities presents a problem for groups that are demanding enforcement of regulation on facilities already in existence;
- It is dangerous to put capital-intensive technologies into unstable countries (safety, proliferation), no matter the degree to which the scope of IAEA operations are increased;
- Present nuclear regulations are both inadequate and insufficient.

Dr. Scott Denman (Safe Energy Communication Council) who also served as an **annotator** to this panel although he was not able to be present for the panel presentation, also provided an anti-nuclear perspective. His main points are:

- On economic, safety, and environmental bases, nuclear energy cannot meet future energy needs.
- Nuclear energy has priced itself out of the market place; and cheaper, cleaner, and safer alternatives exist.

(The basis for these claims on the non-competitiveness of nuclear power was provided by SECCs "Myth Buster #9" document.)

Notes from the authors:

The summaries ended with remarks from two participants who do not see nuclear power as an acceptable component of a sustainable energy portfolio. This is ironic, considering that the overwhelming majority of the preceding dissertators had argued that no mitigation of the greenhouse global warming trends is possible without an important nuclear contribution.

Further, those dissertators had argued that the so-called "four nuclear energy nemeses" – safety, waste, proliferation, and economics – were not real obstacles beyond technical solutions, but artificial constructions that derived strength from public misinformation.

Those dissertators were technical experts; they knew their subject, and they knew they had the right technical answers. When "preaching to the choir," the arguments of these experts appeared so solid as to be self-evident. Still, those arguments did not much to convince the last two annotators.

There is indeed irony here. In fact, before we left the Crown Plaza, we became convinced that this irony was the real lesson to be drawn from this great Conference.

SUMMARY REMARKS: DISCUSSIONS DURING THE CONCLUSION OF THE CONFERENCE

Edward Teller
Hoover Institution - Senior Research Fellow
Stanford University
Stanford, California

There are two obvious reasons why nuclear energy sources must be emphasized. One is that in the not-too-distant future, let us say, the year 2100, the world's energy demand is apt to be at least ten times the present energy production, and presently known methods will not suffice with the only exception of nuclear energy. The second reason is that present worries about harmful environmental consequences are almost certainly justified if present methods are used and the amounts are increased by a factor ten.

The bulk of the papers at this Conference considered in the first place nuclear reactors as they exist today and in the second place, the possibility of thermonuclear reactors. The former approach probably is insufficient. The latter will certainly take time, and in my judgment, will never become competitive.

What is lacking is concentrated work on improvements of fission reactors. Without such improvements, nuclear reactors will not meet the demands in 2100. Some of the possible improvements calling for investigation are as follows:

- Breeder reactors among which I particularly favor fast thorium breeders.

- Automatic regulations minimizing the possibility of human error.

- A thorough re-evaluation of hazards from radioactivity. I recommend that in this evaluation comparison with the natural background be emphasized.

Environment and Nuclear Energy
Edited by Kursunoglu *et al.*, Plenum Press, New York, 1998

- General simplification of operations by eliminating reprocessing and by leaving the long-term radioactivity due to the reactor operation in the original place. This necessitates long-term operations at the same location.
- Appropriate locations of the reactors underground are of real interest in order to reduce health hazards and the danger of proliferation.

The consideration of all these questions should emphasize the conditions under which nuclear reactors could be made available throughout the world without the dangers of accidents and misuse.

I realize that proper answers to these questions are apt to be difficult to find and may in the end not be quite satisfactory. At the same time, I advocate the thorough investigation of a program that as indicated above. This requires thorough but not revolutionary improvements.

SECTION I Basic Requirements For An Acceptable Nuclear Energy Future

ENERGY ALTERNATIVES AND GLOBAL WARMING IN THE 21ST CENTURY*

Chauncey Starr
President Emeritus
Electric Power Research Institute
Palo Alto, CA 94304

INTRODUCTION

The topic of this talk, "Energy Alternatives and Global Warming in the 21st Century" was chosen to set a global energy stage for the coming discussions of the potential role of nuclear power in reducing global warming. Much of the professional literature and media commentaries on global warming reduction focus on constraining energy use by political mandates (as in the coming Kyoto conference) and by economic tools such as taxing and subsidizing. An alternative view is that in a free market society, technology options should be the primary tools for addressing the physical issues of global energy and environment. I will try to shed some light on these energy technology options from my viewpoint as shaped by many decades of EPRI experience in fashioning energy technologies for national, regional, and individual purposes. I will not address the important technologies for accommodating to climate change or potentially for climate modification, such as recently suggested by Edward Teller.

During the past half-century, the global energy concerns of the industrial nations and particularly of the US were primarily focused on assuring the continuity of foreign oil supplies, resulting in confrontations such as the Gulf War. In the world of realpoltik this is still a priority objective. However, in the past decades environmental objectives have become popular in the more affluent nations, with sufficient political support to influence national energy policies. And most recently, the upsurge in political and media attention on the uncertain threat of "global warming" has made environmental concerns global.

The basic issue is that the earth's atmosphere is a "commons" that might be despoiled by any nation to the detriment of all nations. Thus today's US concerns arise from global environmental scenarios which are overwhelmingly dependent on the life-styles

*Based on presentation to International Association for Energy Economics, San Francisco, "Energy: Looking Ahead and Thinking Globally", Sept. 8, 1997.

Environment and Nuclear Energy
Edited by Kursunoglu *et al.*, Plenum Press, New York, 1998

and choices of other countries. Thus, these US concerns need to be addressed by the global energy industry with practical proposals, or the industry will be the unwitting victim of populist short-term political forces. I have doubt that any political entity has both the insight and foresight to plot a global energy strategy for the next half-century in view of the world's unavoidable increasing energy demand to support population and economic growth. There several guesstimates of the future modeled from climate, population, and economic simulations. As a conclusion of a 1997 review of such econometric energy models by the MIT Global Change Program, the projections from such models depend greatly on "--what is assumed about economic growth, productivity improvement in energy use, and the relative costs of future technologies --".

As many of you know, I have a pro-nuclear bias that sees a fear of global warming as an issue that rationally should enhance the public acceptance of nuclear power -- the only major nonfossil source available today. However, I also have a commitment to the use of all energy forms to enhance economic growth worldwide. Nuclear power expansion will require the resources of healthy economies, and their growth will unavoidably depend heavily on fossil fuels. Totally impractical are precipitous and draconian constraints on the use of fossil fuels which must supply the bulk of the world's energy during the next century. We must have the flexibility of multiple options, including nuclear power. We should remember that the real danger is that the public and politicians are often captured by simplistic rhetoric.

Some of you may recall, for example, that such a superficial hype resulted in the US foreclosing natural gas use for power generation by passage of the 1978 Fuel Use Act. Its rationale was scarcity, and its remedy was coal. The Act was repealed in 1987, and now natural gas is our favorite abundant resource. It may seem incredible that such "crystal ball" fortune telling dominants the global warming scene. While this may sound like a curmudgeon's reaction to populism overcoming sweet reason, it describes a pervasive situation in many national energy policies, as evidenced by the US government's treatment of nuclear power. So we should be aware that today's exploration of global energy futures bears the burden of short-term political interests. This turns serious strategies into opportunistic games. I can only hope that the global warming concerns are resolved in a more scientifically reasonable manner, and the important role of nuclear power is recognized.

The Trilemma

The basis of all global energy projections for the coming century is the burgeoning economic growth of the underdeveloped and developing countries. This inevitable growth in global energy demand is added to the more modest projected growth in the OECD countries. As this audience well knows, energy, and particularly electricity, is a keystone to the vitality of modern industrial societies. Thus a global energy scenario requires balancing three fundamental societal goals; economic growth, energy availability, and environment conservation -- called the "trilemma" by our Japanese colleagues (1).

The elements of this trilemma (Fig. 1), (in alliterative terms --population, prosperity, and pollution) are subtly cross-linked by economics, social values, and short-term politics. Unfortunately, the trilemma does not lead to a mathematical optimum to direct global energy strategy. The judgments involved are so dependent on regional cultures, political agendas, and time horizons, that usually only a neutral consensus survives (like the "no-regrets" efficiency policy for global warming). The time dependent interactions between the trilemma components makes their speculative importance variable. They are not independent variables.

Demographers have shown that regional economic prosperity --in the short term-- increases population by reducing infant and old-age mortality rates; and--in the long term-- decreases the fertility rate by reducing the economic value of large families. The empirical finding is that modern industrial societies eventually stabilize at low fertility rates. And similarly with environmental pollution. On the one hand, short-term industrialization and economic development increases the depletion of natural resources (such as forests, arable land, minerals, clean water, pure air, etc.) and also results in an increased output of waste. On the other hand, economic development eventually provides the investment for overcoming such degradation by more efficient use of all resources, resulting in reduced resource demand, recycling, pollution controls, etc. Such resource efficiency generally requires the application of energy, usually by electrification.. On balance, the empirical evidence is very strong that electrification can provide significant improvements to the quality of life. *So the target is not reducing energy consumption but rather to encourage its most productive and efficient use.*

In each society and in each time period, a balancing of the trilemma results from the empirical and political negotiations of the people of that society. We now are undertaking a novel global trilemma negotiation, stimulated by fear of an uncertain climate change that might be induced by mankind's energy use. This is particularly difficult because unlike most site-specific negotiations, there is not today an adequate fact base to provide stakeholders with benefit/cost/risk/time-scale projections of alternative choices. It is unfortunate that in the world of realpolitiks, the narrow self-interests and single-issue demagogues distort the public perception of the trilemma, thus sabotaging a rationale balance -- but we need to keep trying for one.

Energy Technologies

Regardless of climate change, we should seek a mix of foreseeable technologies that might flexibly respond to future shifts in the trilemma balance. While today's mix may be adjusted by future technical improvements, it generally takes many decades to alter a technology's competitive commercial status, so it is instructive to project roles and constraints with today's global mix, assuming continuing trends in growth of population and economics regionally. Such a simplified scenario, based on modest growth rates, was published by Starr and Searl in 1990 (2) and replicated with minor variations by others since. It's summation is illustrated in Fig. 2, and provides us a rough base-case for speculative purposes.. This may be modified somewhat by a gradual increase of the natural gas fraction.

As projected in this scenario, by the middle of the coming century, trends alone lead to a global energy demand roughly 4 times the present. A full effort at conservation might cut this to 2 times present demand by the wide application of known technologies to reduce energy consumption per unit of output. This scenario also indicates an electricity demand increase of 7 times present without conservation, and with full efficiency reduced to an increase of 4 times present demand. Most of this will take place in the developing world. Even in the industrial countries electricity demand will increase. The pressure for growth in per capita electricity use is illustrated by EPRI's scenario for the US, as shown in Fig. 3. The numbers are less significant than the inevitably of such large demand increases. Only a halt to global economic growth, or an apocalyptic population destruction can moderate such demand increases. Supply growth may be constrained by the global capital investment required.

The key message of this scenario is that productive efficiency is the most effective way to reduce global energy demand, and thus the environmental consequences of energy use. Although the capital required for efficiency investment is large (almost the same as investment in supply) past experience suggests that the indirect economic benefits of improved productivity usually makes this a wise economic investment. However, in the short-term, capital for efficiency investment competes with capital for increasing energy supply, particularly in developing countries. It is politically easier to support the supply side of the system rather than the demand side which requires end user cooperation.

A second message is that even at best, global energy use is likely to increase in the next half century to at least double today's. With today's fuel mix, this would mean doubling annual CO_2 emissions, even with the full contribution of all nonfossil sources to the extent that they are physically and economically usable. As we would expect, the environmental movement has been enthusiastic about renewables such as solar, wind, biomass, and occasionally hydro, although their disdain for commercial nuclear power can only be considered as disingenuous. Unfortunately, all the renewables face practical barriers. Hydro is obviously limited and has ecological constraints. Biomass involves energy costs of transportation that generally limits its value to about a 25 mile collection radius around the power plant. The intermittency of solar and wind (diurnal availability about 15-30% in the temperate zone) limits their contribution to peaking or intermittent supplements. Adding energy storage for a continuous base load supply multiplies their capital investment by a rough factor of ten of more, making them economically impractical for such use. Nuclear power is the only non-carbon electricity source that can practically meet the bulk of future global demand.

Generally, electricity supply strategies will be primarily based on proven technologies, and will be chosen in a framework of available capital (domestic and international), social and political stability, and national security. Cost competition will maintain the dominance of fossil fuels for decades to come, even with environmental constraints, but competition among fossil fuels and with nuclear will be intense. In spite of the past difficulties with the first generation commercial plants in the US and elsewhere, nuclear power will have a growing role in countries where long-term capital intensive investments are financially secure and the delivery of oil, coal, and gas is costly or not secure. It is not generally recognized that the new commercial nuclear stations are capital cost competitive with conventional base-load fossil fuel plants, and have the lowest cost fuel with the most secure supply. Only the advanced gas turbine plant is superior to all in gross electricity costs because of its very high conversion efficiency and low capital cost, which overcomes the relatively high cost of pipeline natural gas. In China the competition between nuclear power and pipeline gas will be slowly played out. In Japan and Korea, it will be nuclear power vs. imported liquefied natural gas. Clearly, these are country specific situations, and very technology dependent.

Policy Issues

The inevitability of an increase in annual CO_2 emissions globally is a reality that must be factored into serious discussions of all long-range energy scenarios. Nevertheless, reducing the rate of increase of emissions seems desirable as this extends the time available for accommodating to whatever climate change emissions may induce. For example, a preliminary study by Karl Knapp (3) suggests that a rapid and costly shift to nonfossil electricity generation and nonfossil auto transportation might result in buying a few decades delay in mid-century atmospheric CO_2 levels. *This raises two policy questions. What level of sacrifice today should be made to obtain a few decades delay of an uncertain threat a half century ahead? What would we do with the such added time?*

Although I recognize that many environmental movements have historically opposed nuclear power, I am surprised that they have not actively urged international programs to promote energy efficiency in the developing world. This can have short-term effects, utilizing demonstrated technologies. Of course, such programs imply that modern industrialization and economic growth are worthwhile objectives, and they implicitly acknowledge the inevitability of global electrification and growth in electricity demand. This may be ideologically in conflict with the deindustrialization goals of some environmentalists. For example, we have the rather amazing case of Sweden today, recently studied by Nordhaus (4), where the anti-nuclear Green party is pressuring the state to abandon a low-cost nuclear electricity supply and return to higher cost fossil fuels -- with a consequent sacrifice in global warming goals and economic growth, all against the majority wish of the Swedish public. Fortunately, energy issues are less confused in the developing world where poverty and deprivation are unquestioned environmental targets.

An effective mix of global and regional strategies requires free access to all technologies. I am concerned that suggested government manipulation by fiscal devices, such as taxation and subsidies, would distort the optimal mix that a free technologic competition could sustain over a long-term. For example, a carbon tax intended to reduce CO_2 emissions would obviously distort the free market mix. It would be a self-inflicted harm if limited to the US. As a technologist, I view selective taxation or selective subsidies as a subtle form of censorship, and as a meddlesome hindrance to R&D in today's rapidly shifting and relatively free market of technologies. This should not be confused with government funding of long-term pre-competitive science and technology, which I support. As a global warming example, research on enhancing nature's terrestrial and oceanic CO_2 sinks appears promising and contributes to a common knowledge base. Commercial competition is a different playing field, best left unfettered.

In view of all this uncertainty, it appears to me that the core of any long-range energy strategy is maintenance of the institutional and technical flexibility needed by a globally dynamic energy structure. As a corollary, the major industrial governments have a global responsibility to sustain the long-term viability of all energy options and advanced technologies, specifically including nuclear. This is beyond the economic time span of the commercial sector. New and improved technologies provide opportunities to beneficially fashion the future, rather than only to remedy unwelcome events. Obviously this is a technologist's maximum flexibility "no regret" policy, so I recommend it enthusiastically.

REFERENCES

1. "TRILEMMA, Three Major Problems Threatening World Survival", Susuma Yoda, Ed., Committee for Research on Global Problems, Central Research Institute of Electric Power Industry, Tokyo, Japan. 1995

2. "Global Energy and Electricity Futures: Demand and Supply Alternatives", Chauncey Starr and Milton Searl, Energy Systems and Policy, v.14, p.53-83, 1990

3. "Exploring Energy Technology Substitution for Reducing Atmospheric Carbon Emissions", Karl Knapp, to be published.

4. "The Swedish Nuclear Dilemma: Energy and the Environment", William Nordhaus, Resources for the Future, 1997

THE NEED FOR NUCLEAR POWER

Hans A. Bethe

Floyd R. Newman Laboratory of Nuclear Studies
Cornell University
Ithaca, New York 14853, USA

Nuclear power is not doing well in the USA. No new plant has been ordered since 1973. Many orders placed before 1973 have been cancelled.

By contrast, in France nuclear power has done well. Some 80% of electric power is generated from nuclear fuel. Japan is increasing its nuclear power production steadily. China and other countries in East Asia have big programs to build nuclear power plants.

At present, there is a strong reason for increasing our investment in nuclear power: It is the only known power source which does not contribute to global warming. Other such sources may be developed in the future, for instance solar power. But at present, solar power for large-scale electricity production is prohibitively expensive in most locations. Only nuclear power from fission is sufficiently developed to predict its cost.

About 40% of the energy consumption in the US goes for production of electricity, and this fraction is increasing. At present, only 20% of electric power is generated from nuclear fuel. If we could raise this fraction to 80%, as in France, we would reduce our fossil fuel consumption by about 25%. Our government is committed to a reduction of fossil fuel consumption. I do not see any other way to achieve such a great reduction as 25%.

Burning coal causes air pollution, often also acid rain. Nuclear power is clean, producing only water vapor which soon dissipates.

What about radiation? Everybody agrees that large doses of radiation are injurious to health and can lead to death. But for small amounts of radiation, the evidence is surprising: The cancer death rate in seven western states in the US is 15% lower than in the rest of the continental US, although the level of radiation, both from underground and cosmic rays, is almost twice as high in the western states as in the 41 other states. Studies of homes in China, Japan and the United States that have radon levels up to five times above average indicate that people in those homes have *lower* cancer death rates. There are several other statistics with similar results.

Conclusions of experts differ. The National Council on Radiation Protection (NCRP) believes that the damaging effect of radiation is simply proportional to the total amount of radiation received. But the United Nations Scientific Committee on the Effects of Atomic Radiation (UNSCEAR) declared in 1994 that radiation below a certain threshold has no damaging effect at all. On either assumption, a nuclear power plant next to your home has not noticeable effect on your health.

Environment and Nuclear Energy
Edited by Kursunoglu *et al.*, Plenum Press, New York, 1998

What about accidents? There has been one major accident in a Western reactor; Three Mile Island. The reactor was totally destroyed; a loss of a billion dollars or more to the utility. But hardly any radioactivity was released to the public. It was carefully measured, and even on the pessimistic assumption of NCRP there is just 50% chance that a *single* person near the reactor will die of cancer due to the accident.

Chernobyl, in the Ukraine, was very different. The radioactivity released was about a million times that from Three Mile Island. 28 persons died from acute radiation injury, including a number of firefighters who had flown small planes over the burning graphite of the reactor to douse its fire. If you calculate the cancer deaths by the linear hypothesis of NCRP, you would expect 24,000 cancer deaths, distributed over a population of millions. Fortunately, most of the individual exposures were under 10 rad, a dose which at Hiroshima did not cause statistically significant cancers.

The Chernobyl reactors had a faulty design: If for some reason the power increased enough to drive away the cooling water, the reactivity increased and the reactor became uncontrollable. This flaw has subsequently been corrected in other reactors of the Chernobyl type, by increasing the concentration of U-235 in the fuel. On the day of the accident, the reactor was in the charge of an electrical engineer who was ignorant of the nuclear sensitivity of the reactor. It is very unlikely that any similar accident will occur again in a Russian reactor, and it cannot occur in any reactor licensed in any Western country.

Many people are afraid of the disposal of waste from nuclear power plants. It is most unfortunate that no permanent waste disposal site has yet been established in the US. However, the plans for disposal are very sound. The waste will be enclosed in multiple containment which is not likely to break for many centuries. By that time, the radioactivity will have declined to a very small fraction of the original. It will still be buried inside the rock of Yucca Mountain. The only way it could get back into the biosphere is by transport in ground water. But the disposal site is many hundreds of feet above the ground water level. Only the occasional, very rare rain storm which may seep down as far as the disposal site, and then on down to ground water level, could take small amounts of radioactivity to the ground water. That water has been measured to flow extremely slowly. If any plutonium were dissolved in it, it would constantly be deposited in the surrounding rock and re-dissolved so it would flow far more slowly than the water. The radioactivity of the plutonium and other elements is expected to have decayed long before it could get back into the biosphere.

Deployment of nuclear power will ultimately depend on its cost. The cost of operating these plants looks very favorable: it has declined for the last 10 years and is now on the average about 2 cents per kilowatt-hour, the same as coal-fired plants and much lower than natural gas and especially oil.

The cost of building nuclear plants was very erratic in the 1980s because the Nuclear Regulatory Commission constantly added new safety requirements during the construction of the plants. This has now stopped, the requirements have stabilized, and a number of new designs by the big US manufacturers have been given general approval by the Commission. It is likely, therefore, that the cost of these plants can be predicted fairly reliably in advance.

The running cost for interest and amortization is easily calculable. Assuming 7% interest and a lifetime of the plant of 30 years, a capital cost of $1000 per installed kilowatt translates to about 1 cent per produced kilowatt-hour. Assuming the capital cost to be equal to the median in the 1980s, namely $2000, this would be 2 cents per kilowatt-hour, surely not excessive.

In the near future, many nuclear plants will reach an age of 30 years. Experience has shown that most of them can last much longer. If some of them have to be retired, it is important that they be replaced by nuclear, not by fossil fuel plants, so that at least we do not lose further in our fight to reduce global warming.

NEW DIRECTIONS IN THE DEVELOPMENT OF NUCLEAR POWER

Glenn T. Seaborg

Today there is a strong movement in the direction of developing nuclear power that is more acceptable to the general public. I have had a front seat in the development of the water-cooled reactors that are producing nearly 20% of the electrical energy being developed in the United States. I believe that these reactors are relatively safe compared to other methods of producing electricity today – in fact, they probably represent the safest method of producing electricity. However, that is not the topic of my essay, which is devoted to new developments in the production of nuclear electric power.

I shall describe briefly these new directions in the development of nuclear fission power and then go on to nuclear fusion power.

Presently, in the United states there is an aggressive program to develop and standardize nuclear fission power reactors that take advantage of some 35 years of experience. There are several types of such nuclear power systems under development and/or operating. The Advanced Boiling Water Reactor (ABWR), pioneered by General Electric, is now operating in Japan. The System 80+, an advanced Pressurized water reactor is being developed by ABB Combustion Engineering based on its positive experience with the System 80. And the AP600, a small "passive" PWR is under development by Westinghouse. One should note that both the ABWR and the System 80+ have been given "standardized" licenses by the Nuclear Regulatory Commission (NRC). The AP600 is currently under NRC review.

A point to be made is that nuclear reactors built to U.S. standards have been absolutely safe for the public worldwide since the start of the peaceful nuclear era in 1954, and have demonstrated their practicality despite the finding of a number of problems over the years. The new reactors are intended to correct the problems and incorporate the positive elements developed over the years, and to reduce capital and operating costs. They appear to be accomplishing these objectives, although they do not appear to be competitive with gas fired fossil plants at today's low gas prices. The price of gas may rise in the future; and/or be taxed for its CO_2 emissions.

Until March 1995 the U.S., through a Department of Energy supported program, was intensely working on the development of the Liquid Metal Reactor (LMR). The LMR (a Breeder Reactor) can obtain 50 to 100 times as much energy from a pound of uranium than do our present type of reactors. In

Environment and Nuclear Energy
Edited by Kursunoglu *et al.*, Plenum Press, New York, 1998

the next century when uranium supplies for our present reactors may become scarce, the LMR may be vital in supplying needed energy to the world. Thus, I suggest that U.S. Policy on the LMR be reconsidered; last year the U.S. LMR program was canceled, although it continues abroad.

The U.S. Liquid Metal Reactor (LMR) program was focused on the PRISM (Power Reactor Inherently Safe Module) concept. The PRISM concept utilizes four identical reactor modules arranged in two 800 Mwe power blocks for an overall net electric rating of 1600 Mwe. The reactor itself is a pool-type sodium-cooled reactor with an intermediate heat transport system to isolate the steam generators from the reactor in the event of a steam generator leak and subsequent sodium/water reaction. Simplicity and inherent safety have been build into PRISM in the form of passive shutdown heat removal, passive accommodation of Anticipated Transient Without SCRAM (ATWS) events, and passive post-accident containment heat removal. PRISM has received very positive reviews from the NRC during its development.

The LMR and the present type advanced reactors mentioned above have the advantage, as do the present conventional fission reactors and the fusion reactors that I will describe, that they circumvent some of the problems accruing from fossil-fueled electricity generating plants – acid rain, the greenhouse effect, the tremendous health toll on coal miners, and the ultimate depletion of fossil fuel supplies.

Now I turn to the role of nuclear fusion reactors. Hopefully, the present types can be developed for economic operation, if this should prove possible at all, on an unfortunately long time scale – that is, the magnetic confinement and the inertial confinement fusion reactors. Although such reactors do not have the fission product waste disposal problem attendant with fission reactors, they are accompanied by neutron emission leading to large quantities of radioactive byproducts, and they are also plagued by the difficult problem of converting this neutronic energy into electric power. Required here are extremely high plasma temperatures or difficult laser implosion techniques that have defied the ingenuity of investigators for many decades, with the prospects that such hurdles may not be overcome for many decades to come, if ever.

We have all been intrigued by the reports from Utah in 1989 suggesting that "cold fusion" has been observed in the electrolysis of heavy water, utilizing a palladium electrode. The relative yield of neutrons was reported to be very low, which may not be an advantage, depending on the validity of the experiments. These observations caught the interest of scientists in some laboratories in this country and abroad with the result that much doubt has been cast on whether this can ever be a practical source of energy or even whether nuclear fusion reactions have actually been observed. I shall not comment further on this.

I shall comment on a unique approach to nuclear fusion – an approach that could lead to aneutronic power, essentially nonradioactive nuclear energy. This involves the concept of colliding beam fusion, featuring such reactions as $D + {}^3He \longrightarrow H + {}^4He$ and ${}^{11}B + H \longrightarrow 3\,{}^4He$, leading to aneutronic charged particle products. The energy released in such aneutronic reactions is in the form of charged particles which can be converted directly to electricity with a very high efficiency. No other form of energy has ever come close to this degree of cleanness and efficiency. In all power sources today approximately two-thirds of the energy is in the form of waste heat or heat pollution. We may apparently be at the brink of an energy technology which is compact, safe, and has virtually no residue but the electricity we see fit to create with it. There is virtually no waste at all and none of it dangerous. It is conceivable that this technology could be miniaturized, revolutionizing electrical transmission distribution and decentralizing energy production.

The greatest energy needs are those of the developing world. There can be no development of industry without a new source of energy. The spread of nuclear energy to the developing world has been prevented by two factors: (1) high capital cost of nuclear reactors and (2) non-proliferation regulations.

How could the American innovative capacity be harnessed to develop a nuclear power system that the world can use? There is a large potential demand for a small modular nuclear power plant (5~100 Mwe), if it were able to be built with low capital cost and was inexpensive, mass produced, proliferation-proof, and radiation-safe.

Neither the industry nor the government has thus far made more than a trivial study of the possibility of developing the power plant that would meet these market demands. The projected aneutronic power plants promise to be small and modular, thus cutting the capital cost and since they are neutronless, they cannot proliferate. I am not the first to declare that this would "meet the original goal of the fusion program: universally available, inexhaustible, environmentally benign power."

The planet which we inhabit is in the midst of a serious environmental crisis. In large part this has been brought about by our efforts to meet the constantly growing demand for energy. It is now apparent that lack of energy resources is not the problem, but the threat to our environment from the unused residue of energy production. We are in an energy residue crisis, and this is how we must look at energy technology from now on. We cannot allow the unused residue of our energy production – waste heat from thermal conversion, carbon dioxide from fossil fuels, radioactive waste from nuclear fission power – to destroy the environment while we accommodate our needs. We must find solutions to these problems.

I was much involved in the development of nuclear fission engineering, and have shared the excitement. I know well what a disappointment it is to experience the difficulties and frustrations in coping with the environmental hazards of fission reactors. But being aware of the enormous challenges we have overcome to get where we are, I am all the more convinced that we must not lose sight of the need to develop clean and safe nuclear power.

It has been only 100 years since the discovery of radioactivity, and some 60 years since the discovery of nuclear fission. In historical terms, this is an extremely short time. We may just have began to explore the world of nuclear energy. We are still pioneers. Nuclear engineering has dared greatly, and encountered great obstacles. But we may be at the beginning, not the end, of the nuclear age.

NUCLEAR ENERGY FOR THE THIRD MILLENNIUM

Edward Teller
Hoover Institution - Senior Research Fellow
Stanford University
Stanford, California

INTRODUCTION

The major energy sources of today are expected to last for only a small fraction of the millennium starting in the year 2000. In the plans of most people, nuclear energy has been ruled out for four separate reasons:

1. The danger of radioactivity from a reactor accident or from reactor products during a long period after reactor shutdown;

2. The proposed fuels, U-235 and also Pu-239, as obtained by presently available procedures will serve only for a limited duration;

3. Energy from nuclear reactors will be more expensive than costs of present alternatives;

4. The possibility of misusing the products for military purposes is an unacceptable danger.

The development described below attempts to meet all four objections. Specifically, we propose[1] a structure as an example of future reactors that is deployed two hundred meters underground in loose and dry earth. The reactor is designed to

[1] This proposal has been presented in 1996 in Obninsk, Russia by Teller, Ishikawa, Wood, Hyde and Nuckolls under the title ***Completely Automated Nuclear Reactors*** in the Proceedings of the International Conference on Emerging Nuclear Energy Systems (ICENES '96), and is available at http://www-phys.llnl.gov/adv_energy_src/ICENES96.html

Environment and Nuclear Energy
Edited by Kursunoglu *et al.*, Plenum Press, New York, 1998

function for thirty years, delivering electrical power on demand up to a level of thousand electrical megawatts. From the time that the reactor is started to the time of its shutdown thirty years later, the functioning is to be completely automatic. This is an obviously difficult condition to fulfill. The most important factor in making it possible is to design and operate the reactor without moving mechanical parts.

At the start, the reactor functions on thermal neutrons within a structure containing uranium enriched in U-235 or having an addition of plutonium. That part of the reactor is to deliver energy for approximately one year after which a neighboring portion of the reactor containing thorium has been converted into Th-233 which rather rapidly decays into fissile U-233. This part of the assembly works on fission by fast neutrons. It will heat-up if insufficient thermal energy is withdrawn from the reactor's core due to the action of specifically engineered units. Indeed, these specifically designed units absorb neutrons (while releasing little energy) if excessive reactor co-heating occurs. These units played the role of automatic control rods and will be described below.

After the thorium in a given volume of the reactor's fuel charge is depleted, an adjacent thorium-containing portion of the fuel charge will have been converted—bred—into fissile material and is ready to continue the reaction. We may think of the system as a series of neighboring reactors where each reactor, while exhausting its fuel, activates its neighbor. An actual practical plan will be to merge these reactors rather than construct them from separate units. Thus, one may obtain propagation of burning from one end of a line to another or from an initial central position to two ends of a line.

After the thorium in the reactor's fuel charge has been used up, the reactor is shut down by the first positive action of the operators in thirty years. The residual radioactivity will be sealed within the reactor's core and thereafter allowed to decay in place. The initially intense radioactivity will leave the reactor products inaccessible and unusable for military purposes except if complicated, expensive and easily observed large-scale operations are performed. Having thereby avoided transportation of fission products and reprocessing thus significantly reducing cost and hazards.

CONVERSION OF HEAT INTO ELECTRICITY

As stated above, production of nuclear heat in this new type of reactor system occurs at an underground depth of approximately 200 meters. In most present reactor designs, the heat is transferred to units generating electricity by the means of steam. We plan to replace water cooling by helium cooling, thereby permitting reactor core operation at higher temperature and higher thermodynamic efficiency. At the same time, helium is chemically inert at all temperatures while water becomes chemically quite active at high temperatures.

The generation of electricity is to occur above the surface or perhaps slightly underground. The hot helium is to be used in a manner similar to the functioning of modern combined cycle generating units burning natural gas. The rate of power generation is regulated by the rate of pumping hot helium out of the reactor's core.

For long-term addressing of energy supply, it is essential to burn-up a major fraction of the readily available actinides, e.g., thorium. This means that we must utilize the methods discussed under the designation of "fast breeders" (which refers to fast neutrons carrying the nuclear chain reaction). Indeed, slow (thermal) neutrons are strongly absorbed by fission products, so that any candidate slow breeder could utilize only a small fraction of thorium (or uranium) before the accumulation of the fission products would prevent further thermal neutron chaining.

The functioning of a fast breeder is practically independent of temperature and, therefore, means must be arranged to function as "control rods" in order to quench neutron multiplication at excessive temperatures. One means by which this can be accomplished is by constructing thermal-neutronic units containing the readily available light isotope of lithium, Li^6, which is a strong neutron absorber. Many small reservoirs of lithium-6 are located outside the core of the reactor, where the residual neutrons contribute little to the power-producing chain reaction. Small modules of normal lithium are positioned throughout the reactor's fuel charge. When the reactor temperature becomes excessive, thermal expansion of the normal lithium pushes Li^6 from out of the reservoirs through capillary tubes into compartments within the fuel where the neutron absorption by Li^6 will quench the chain reaction. Lithium is a liquid throughout the temperature range of reactor operation. The actual development of this lithium-controlled-rod is a technical undertaking which is yet to be worked out. By redundantly introducing these units in large numbers throughout the nuclear fuel, we practically produce a very strong negative temperature coefficient of reactivity above the design temperature and insure that by varying heat withdrawal from the core, we can regulate the reactor's power output. If the coolant He moves rapidly through the reactor's core, the Li^7 will contract, Li^6 will be withdrawn from the core, and heat production will be stimulated. If helium moves slowly, core temperature will rise and heat production will be quenched.

We expect that there may be challenges in realizing mass-producible Li thermostating elements for highly reliable thirty-year operation. In particular, it is important that the expansion and contraction of Li^7, as well as the associated flow of Li^6, should remain quickly and thoroughly reversible, even after thousands of operations. However, the needed elements could be developed and tested in high flux, slow neutron regions of an existing reactor. It is to be noted that operator-governed mechanical motions of control rods are replaced by a fully automatic, highly redundant arrangement that depends simply on thermal expansion.

This automatic and redundantly applied arrangement is one of several in this reactor system design which aims at reducing the probability of any accident. But we expect, that even in case of an accident, human health will not be endangered.

It is necessary to point out that even with inclusion of Chernobyl into the overall statistics, lives lost divided by energy generated throughout the world has been far less for nuclear reactors than for all conventional methods of electricity generation.

Sohei Kondò, in a specific study entitled ***Health Effects of Low-level Radiation,*** states in the Preface on Page xiv:

> *The collected data strongly suggest that low-level radiation is not harmful, and is, in fact, frequently 'apparently beneficial' for human health. The data are incomplete, however, and presentation of these facts alone will not suffice to allay the fear of low-level radiation, especially with regard to the risks for congenital anomalies and cancer.*

Transportation of reactor products to a central location and storage of the products in a limited region may increase hazards. Our present proposal is to avoid the costs of such transportation and storage, and at the same time, decrease hazards by locating power reactors in appropriate regions. The main requirement is absence of water that might transport radioactivity to the surface and thus into the biosphere. If in course of geologic changes, water should approach the ancient reactor site, we calculate that due to the high temperature in this region arising from the slow decay of the reactor's still-contained nuclear waste products, the water will evaporate before it reaches the vicinity of the reactor itself. In planning the locations of these reactors, it will be necessary to give special attention the dry regions of the Earth's surface. At the same time, it also appears to be desirable to locate reactors in relatively loose earth in which earthquakes are not apt to open fissures.

ANSWERS TO THE FOUR OBJECTIONS AGAINST NUCLEAR POWER

1. Numerous nuclear explosions have been performed at depths of roughly 200 meters in Nevada in comparatively loose earth. The energy made swiftly available is often comparable to or more than what could be released in a reactor accident. In spite of this, no dangerous amounts of radioactivity have reached the surface. In addition, useful information was obtained about the slow migration of radioactivity underground. We believe that these results may be applied anywhere else, in comparable conditions.

It follows that the reactor should be so located as to exclude access by water for the present time and as far into the future as possible. While no guarantee can be given concerning the effects of geological changes, the heating of the material surrounding the reactor happens to be a favorable factor. One may thereby expect that invasion by water (unless it is rapid and massive) will be counteracted by its evaporation before it reaches the vicinity of the now-ancient reactor and its residual radioactivity.

2. In our proposal, enriched uranium and plutonium are used for only a few percent of the nuclear heat eventually generated. The great preponderance of the energy developed comes from thorium, and this element is available in practically mineable deposits in quantities sufficient for millennia of present-day electricity generation.

3. Expense of future deployment and operation is hard to predict. The needed excavation is an old and well-developed technology and is not likely to contribute more than 10 to 20 percent of the total system cost.

Long-term availability of the fuel has been already mentioned and such fuel can be used without any expensive intermediate step such as isotope separation.

One of the particularly important innovations that have been proposed is the fully automatic operation, effectively obsoleting human intervention. Because of its unusual character if applied to an operation over many decades, this is a particularly critical issue. One reason why I hope for success is the relatively simple functioning of the energy-producing mechanism which we expect to work without programmed moveable parts. It is this automatic operation that may need most attention in fully developing the proposed reactor technology.

A considerable part of the cost of nuclear power in the U.S. at the present time is due to delays in the licensing of new nuclear power plants. These delays result in money "lost" paying interest on early investment. Standardized design and decrease of fear of accidents should greatly facilitate elimination of such delays and costs.

Even at the present time in the U.S., nuclear power generation is accomplished without a major fraction of the electricity cost being due to costs of operation. In the proposed scheme, the cost of operating the reactor should be much smaller.

A closely-related question is the availability of reliable operators. In worldwide usage including developing countries, fully-qualified operators may be unavailable in quantities required. If, indeed, we can do without them, this would be a great advantage in making this energy source generally available.

4. The connection between energy production and material for nuclear explosives is obviously worrisome. In our proposal, the residual fissile material is left underground and will remain radioactive for centuries. To obtain this material will require automated equipment in massive, sustained operations. The use of such equipment will be complicated and expensive. If arrangements are made to overcome these difficulties, it will require the kind of complicated and large-scale operation that is difficult to keep secret. Thus, the widespread availability of nuclear electricity will be achieved without substantial danger of proliferation of nuclear explosives.

CONCLUSION

The acceptance of our proposal is hampered by the circumstance that rather unprecedented procedures are needed. Modern technology has in general been exercised by ample use of humans in all phases of planning, construction and operation. We are proposing the practical elimination of people from the operational phase. To do this in a reliable manner will obviously engender opposition.

It should be emphasized that three of the four objections against nuclear power are resolved in part by reducing the number of people involved and, if possible, by eliminating them from the operational phase.

The serious accidents of Three Mile Island and Chernobyl were brought about by severe operating mistakes. Simple adherence to existing safety procedures would have avoided both accidents. As long as the public insists on complete avoidance of accidents, we must insist on straightforward safety measures operating in an automatic manner. As has been pointed out, eliminating the unreliability which may be inherent in human operators may be a necessary step if nuclear energy production should have worldwide application.

Most importantly, the automatic operation would make it possible to operate nuclear power reactors under governments which may not be entirely trustworthy.

The peculiar conclusion is that underground reactors are not compatible with imperfect solutions of a number of our problems. But they can have really great advantages if all the problems indicated are solved. The proposed solutions of underground location and breeding are not novel. But the combined effects of these innovations may deserve more attention.

I want to thank Dr. Lowell Wood for his help and for important contributions.

NUCLEAR ENERGY, PAST, PRESENT, AND FUTURE

Gary M. Holahan

Director, Division of Systems Safety and Analysis
U.S. Nuclear Regulatory Commission
Washington, DC 20555

"Atomic energy is capable of applications for peaceful as well as military purposes." With this pronouncement, the Congress, in the preamble to the 1954 Atomic Energy Act, released atomic energy (nuclear energy) from its military origins. The Congress also set out a new purpose for this new energy source...to serve and improve the "general welfare" of the country. The intent of the act was clearly broad and substantial, "to provide for...a program to encourage widespread participation in the development and utilization of atomic energy for peaceful purposes."

Those original goals have been fulfilled only to a modest extent. Medical and a few industrial applications of nuclear byproducts have become a normal part of our modern society. The role of radiation in medical diagnostics and treatments has been accepted, if not welcomed. In some sense, nuclear medicine is seen as no better and no worse than other medical procedures and is clearly preferable to the diseases it addresses. With respect to commercial power plant applications, nuclear energy has not been integrated into the electrical energy production mix of the nation in the same way. The public does not see nuclear power plants as just one more way of producing electricity. The issues are seen as different, special, and of greater concern.

In order to address any potential role that nuclear energy could play in the future, it is first necessary to consider its past and its present, in terms of both our best technical understanding and in terms of public perception and acceptance. I will address these issues from the viewpoint of the regulatory agency; first the Atomic Energy Commission (the AEC) and now the Nuclear Regulatory Commission (the NRC).

Any future nuclear power plants; must be safe and environmentally acceptable by the best technical assessments available (including reactor operation; decommissioning; and waste disposal); must be perceived as safe and environmentally acceptable by the public; and must be economically competitive. I will limit my remarks to issues relating to nuclear reactor operations.

Before the 1960's, nuclear reactors were in an experimental or developmental stage. Fuel performance was uncertain and several partial or total core failures were experienced.

Environment and Nuclear Energy
Edited by Kursunoglu *et al.*, Plenum Press, New York, 1998

Accidents at EBR1 (1955), Chalk River (1952 and 1958), SL1 (1961) were among those occurring at experimental reactors. This experience is not unlike the early experience in other developing technologies. Demonstration projects tested new approaches: pressurized-water reactor (PWR); boiling-water reactor (BWR); sodium-cooled reactor; gas-cooled reactor; and organically cooled and moderated reactor. Some worked well, others didn't. The public was largely unaware and uninterested. Government institutions and activities were generally trusted and supported.

However, by the early 1960's a first generation of small commercial nuclear power plants were being put in place: Dresden 1, Big Rock Point, Humboldt Bay, Indian Point 1, and Yankee Rowe among others. These plants were all small by modern standards. They also lacked a common set of safety standards. The major safety issue of the 1960's became the need for a reactor containment building and the primary safety issue of the early 1970 became the need for an emergency core cooling system (ECCS). New rules were written and new safety standards put in place. The General Design Criteria (10 CFR Part 50, Appendix A) are the most conspicuous example of those safety standards. With these new and clearer safety requirements in place for core design, thermal hydraulic performance, ECCS capability and containment capability, the reactor designers began to develop larger and larger plants, constrained more by the industrial capability of manufacturing large components than by any safety or regulatory considerations. With no clear end in sight for proposed power levels, the regulatory agency set a somewhat arbitrary limit corresponding to about 1300 MWe as a maximum licensable power level. At that point, five U.S. companies were actively designing nuclear reactors and numerous utility companies were eagerly buying them. While these "maturing" steps were being taken in the area of nuclear power plant design, the nation as a whole was loosing its confidence in government and corporate institutions.

In hindsight, with a fuller appreciation of: probabilistic risk assessment techniques; of operating experience; and of human performance, the risks from operation of these power plants seems substantially higher than for current plants, yet both the government and the public were more accepting of them.

In 1974 the first comprehensive Probabilistic Risk Assessment for nuclear power plants was published by the AEC in the "Reactor Safety Study" commonly called "WASH-1400." It presented the first full picture of the risks of operation for a typical PWR and a typical BWR. Although criticized as "inscrutable" and incomplete, WASH-1400 presented reactor safety insights in a scientifically consistent and coherent manner. For the first time the designers, the operators, the public, and the regulator could see what was truly important and what was unimportant with respect to public health and safety. This was the first opportunity to introduce a regulatory approach which was more directly related to perceived risks. That opportunity was missed. The combination of: technical criticism; immaturity of techniques; limited operational data; a focus on plant licensing; and operational events put WASH-1400 in the background.

In 1975, the worst fire in U.S. nuclear power plant experience occurred at the Browns Ferry plant. That fire pointed out a significant weakness in their existing regulatory requirements. Resolution of the fire protection concerns resulted in several years of contentious arguments between the licensees and the NRC. The resulting fire protection rule (10 CFR 50.48) resulted in a significant improvement in fire protection but has been seen as overly prescriptive (i.e., inflexible) in its approach and not completely focused on addressing public health and safety issues.

On March 28, 1979, the TMI-2 accident occurred and the future of nuclear power in the U.S. would be changed forever. The messages from that accident were clear and unmistakable: 1) severe accidents leading to core damage were a real concern, and could no longer be treated as remote, hypothetical, and unrealistic; 2) the importance of the defense-in-depth philosophy was re-confirmed; 3) neither the licensees nor the NRC was prepared to respond effectively to a severe accident; 4) the human factors and human performance aspects of reactor safety were very important yet had been largely overlooked; 5) the actual operating experience of nuclear power plants had been ineffectively monitored by both the NRC and the nuclear industry. The TMI Action Plan of over 170 new requirements was synthesized from the numerous accident studies. In addition, the NRC's Commission Policy Statements on Safety Goals and on Severe Accidents were direct attempts to recognize the importance of severe accidents and to address them in several practical ways.

One of the most significant of the TMI-2 accident studies was the so called "Rogovin Report" initiated by the Nuclear Regulatory Commission. In addition to its direct contribution to the insights listed above, the Rogovin Report represented the second opportunity to introduce Probabilistic Risk Assessment (PRA) techniques directly into the regulatory process. Its recommendations in this area are clear.

"More rigorous and quantative methods of risk analysis have been developed and should be employed to assess the safety of design and operation.

The best way to improve the existing design review process is by relying in a major way upon quantative risk analyses..."

This second opportunity was only partially fulfilled and only after several years. In 1988 the NRC issued a request to the licensees of nuclear power plants to voluntarily perform a self assessment (the "Independent Plant Evaluations" (IPE's)) using PRA techniques (or their equivalent). As valuable as the IPE's have been, they were not the change in regulatory approach envisioned in the Rogovin Report.

The 1980's were largely a reactive time in which safety improvements were required in numerous areas. The safety requirements added to plants in the years since their original licensing (i.e., in the 1970's, 1980's, and early 1990's) include: the TMI Action Plan items; the ECCS rule (10 CFR 50.46); the Emergency Planning Rule (10 CFR 50.47); the Fire Protection Rule 10 CFR 50.48); the Electrical Equipment Qualification Rule (10 CFR 50.49); the Anticipated Transients Without Scram (ATWS) Rule (10 CFR 50.62); the Pressurized Thermal Shock (PTS) Rule (10 CFR 50.61); and the Station Blackout (SBO) Rule 10 CFR 50.63). In addition, numerous other requirements were put in place through other safety initiatives in response to operational events.

As a time of increasing requirements, increasing costs, and slowing growth in electrical energy demands, the 1980's were not encouraging to utilities thinking about adding electrical production capacity. The utilities which were operating nuclear power plants concentrated on improving operations and have been very successful at increasing capacity of their plants while reducing the number and severity of events. The nuclear power plant designers became largely service companies for the short term and began to look toward standard evolutionary or advanced designs for the more distant future (and for foreign markets).

And where is nuclear power now in the 1990's? Operating plants are generally performing better than at any time in the past as measured by the NRC Performance Indicators or by those of the Institute for Nuclear Power Operations (INPO) but some individual licensees performance has been of concern and has called for additional regulatory attention. In 1996, there were 5 U.S. plants among the top 10 nuclear power plants in the world with respect to overall capacity; and 23 among the top 50. Yet economic pressures has forced several U.S. plant closures before the completion of the plant's design life.

The NRC has established clear regulatory processes for nuclear power plant standard design certification and identified increased safety expectations. In the NRC's Strategic Plan (NUREG-1614, Vol. 1), the Commission has established a strategy to "-give priority attention for license renewals, standard and advanced reactor design, early site approvals and new reactor licenses." It has gone on to review and approve two new standard designs (the evolutionary PWR design, ABB/CE System 80+ and the advanced BWR, GE's ABWR) and is currently reviewing an additional advanced design (the Westinghouse advanced PWR, AP-600). With respect to Probabilistic Risk Assessment, the current NRC initiative which NRC's Chairman Jackson has dubbed "Risk Informed Regulation," is the strongest attempt yet to reform the regulatory process to assure that it is focused on those issues most directly related to public health and safety. The NRC recently issued for public comment, ten guidance documents which define the essential safety principles and expectations associated with a Risk-Informed Regulatory approach. Several pilot projects are also underway. As a result of these activities the NRC expects to: improve safety decision making; make better use of NRC staff resources; and to reduce unnecessary burden on NRC licensees.

The current regulatory environment can be summarized as follows. The NRC is prepared to review, approve, and license new designs which have been shown or can be shown to meet modern safety requirements. The NRC is prepared to, and is in fact already at work, reforming regulatory requirements to focus them better on public health and safety issues and to provide added implementing flexibility where and to the extent practical. The NRC has an effective program for monitoring operational performance and will acknowledge good performance and respond effectively to poor performance. The technical safety issues relating to continued operation of current plants and the licensing of future evolutionary or advanced designed can be sufficiently met to allow continued or even increased used of nuclear power. The more difficult issues of: public perception of safety, public acceptance of nuclear options; and economic feasibility are likely to be the real determining factors in any future for the nuclear power option in the U.S. With respect to public acceptance of nuclear power, two conditions are necessary (although they may not be sufficient). They are: 1) Long term safe operation of current plants and 2) credible, open, and independent regulatory oversight which is responsive and understandable to the public.

SECTION II Global Energy Challenges Arising From the Demands Of Industrialization Of the Regions With Large Populations

THE ROLE OF NUCLEAR ENERGY IN MITIGATING GREENHOUSE WARMING*

R. A. Krakowski

Systems Engineering and Integration Group
Technology and Safety Assessment Division
Los Alamos National Laboratory
Los Alamos, New Mexico 87545

ABSTRACT

A behavioral, top-down, forced-equilibrium market model of long-term (~2100) global energy-economics interactions has been modified with a "bottom-up" nuclear energy model and used to construct consistent scenarios describing future impacts of civil nuclear materials flows in an expanding, multi-regional (13) world economy. The relative measures and tradeoffs between economic (GNP, tax impacts, productivity, *etc.*), environmental (greenhouse gas accumulations, waste accumulation, proliferation risk), and energy (resources, energy mixes, supply-side *versus* demand-side attributes) interactions that emerge from these analyses are focused herein on advancing understanding of the role that nuclear energy (and other non-carbon energy sources) might play in mitigating greenhouse warming. Two ostensibly opposing scenario drivers are investigated: a) demand-side improvements in (non-price-induced) autonomous energy efficiency improvements; and b) supply-side carbon-tax inducements to shift energy mixes towards reduced- or non-carbon forms. In terms of stemming greenhouse warming for minimal cost of greenhouse-gas abatement, and within the limitations of the simplified taxing schedule used, a symbiotic combination of these two approaches may offer advantages not found if each is applied separately.

INTRODUCTION

The Los Alamos Nuclear Vision Project[1,2] is investigating a range of possible futures for nuclear energy using the construct of scenario building[3,4] and an established, relatively

* The sampling of results presented herein is elaborated in Los Alamos National Laboratory document LA-UR-97-4380 (October 29, 1997) of the same authorship, title, and textual material.

transparent global energy model.[5] Both nuclear energy demand and the flow of nuclear materials are examined over a ~100-yr time horizon that is characterized by a range of scenario descriptors or attributes [*e.g.*, population growth, work-force productivity (GDP), autonomous energy efficiency improvements (AEEI, or non-price improvements in transforming primary and secondary energy to energy services), energy resource constraints, carbon taxation schedules, capital- and operating-cost constraints imposed on a range of nuclear energy technologies, *etc.*]. While the focus of past analytical support of the Nuclear Vision Project[6-10] has been on issues and concerns related to global implementation of an expanding nuclear fuel cycle, the "top-down" behavioral model of an equilibrium (*e.g.*, market-clearing, supply = demand) energy market embodied in the ERB (Edmonds, Reilly, Barns)[5] model adopted and modified for this study also delivers estimates of greenhouse-gas (GHG) emissions. Hence, coupled with the "bottom-up" nuclear energy model[6,7] that has been matched to the recursive, "top-down" formalism of the ERB model, with this nuclear model providing regional and temporal tracking of plutonium inventories and forms and a relative measures of nuclear proliferation risk[10] based on earlier work,[11-14] top-level energy/economic/environmental (E^3) trade offs become possible.[7,9] Furthermore, by implementing (into ERB) integral-response functions[15] that have been calibrated against a global atmospheric-ocean climate-change model,[16] the GHG emission rates reported by ERB for an array of scenario attributes can be expressed in terms of atmospheric CO_2 accumulations, W(GtonneC), and increases in average global surface temperature, ΔT(K). Within the limitations of the modified ERB model and with little additional effort, the role of nuclear energy in mitigating greenhouse warming can be examined under the above-mentioned scenario construct, with all three of the Es in E^3 being touched at some level.

Nuclear energy, like solar and (equilibrated) biomass energies, is a non-carbon (NC) energy source that has clear GHG-mitigating potential. The role played by non-nuclear NC energy sources is limited to the economic constraints that form the basis of the original ERB model,[5] although recent studies of the GHG-mitigating potential of (equilibrated) biomass energy sources has been reported.[17] The present study focuses on the nuclear-energy option, and efforts to consider other NC energy sources in the context of the present effort remain as future work. This present focus on a "bottom-up" nuclear model without comparable examinations of other NC options is a serious limitation. Furthermore, only electricity generation is considered for the nuclear options being considered; since $\gtrsim$ 60% of all primary energy presently serves fossil-based non-electric applications, this too is a serious limitation of the present study. Lastly, mitigation of greenhouse warming through the implementation of NC energy sources attacks the problem only from the supply side. Increased demand-side energy efficiencies represent the other main facet of the problem.[17-20] This (demand-side) approach to GHC mitigation is examined herein through the aforementioned AEEI parameter; in the context of the ERB model, AEEI is changed parametrically (exogenously). More elaborate (long-term) models reflect endogenous increases in either AEEI,[21] if that concept is used, or induced reallocations of resources among key sectors of the world economies as non-energy sectors adjust to increased energy prices.[22] The ERB model is capable only of exogenous changes in the AEEI-like parameter, ε_k. A fourth limitation of this analysis centers on the merits of economy-based "top-down" models *versus* technology-base "bottom-up" models;[23] the former generally reflects market penalties associated with GHG mitigation schemes, whereas the latter solution-oriented (and generally market-free) approach suggests cost benefits for changing to reduced- or non-carbon fuels and using those fuels more efficiently.

With these four limitations in mind (*i.e.*, nuclear-energy-focus; application only to electricity generation; exogenous AEEI; "top-down" approach), the results summarized in Refs. 8 and 24, along with the associated technical support document,[9] are directed at understanding better the role nuclear energy might play in abating greenhouse warming. After giving a synoptic narrative describing the ERB model, results are presented according to the following four subsections: a) description of the Basis Scenario; b) impacts on global warming by exogenously varying AEEI (*e.g.*, demand-side impacts); c) the impacts of increased nuclear-energy share fractions induced by imposing a range of carbon-tax schedules (*e.g.*, supply-side impacts); and d) the composite impact on global warming of simultaneous demand-side (increases in AEEI) and supply-side (imposition of carbon taxes) forces. On the basis of these parametric results, a nuclear-energy scenario that mitigates greenhouse warming is suggested. This "strawman" scenario combines both supply-side (carbon-tax induced increase in nuclear-energy demand) and demand-side (AEEI increases) approaches. An interim summary and conclusions follows.

MODEL

Four basic approaches to modeling energy planning have evolved[25] over the years: a) simulations of the technical system *per se*;[26] b) econometric estimates of price-demand relationships;[27] c) sectoral descriptions of whole economies, with energy being one of a number of interconnected sectors;[22,28] d) optimization models that combine elements of the others into a Linear Programming (LP) or a Mixed Integer Programming (MIP) formulation.[21,29,30] The ERB model uses a recursive approach to describe a behavioral market equilibrium that internally balances energy production and usage. As such, the simplified ERB formulation models energy from within using econometric price-demand relationships. While simplified compared to the sectoral and/or LP optimizing models, the ERB model adequately targets the needs of the present study, is available to the public, is adaptable to modification, and is generally transparent and well documented.[5] While presenting a "top-down" economist's (market) view of E^3 interactions, an approximate "bottom-up" technology view of nuclear energy has been added.[6]

The ERB model is comprised of four main parts: supply, demand, energy balance, and GHG emissions (a postprocessor). Appropriate carbon coefficients (Gtonne/EJ) are applied at points in the energy flow where carbon is oxidized and released to the atmosphere; carbon flows where oxidation does not occur are also taken into account. Supply and demand are determined for six primary energy categories: oil (conventional and nonconventional); gas (conventional and nonconventional); solids (coal and biomass); resource-constrained renewables (hydroelectric and geothermal); nuclear (fission, with fusion being included as a form of solar energy[5,31]); and solar (excluding biomass; includes solar electric, wind, tidal, ocean thermal, fusion, and advance renewables; solar thermal is included as a form of energy conservation). The energy balance in ERB assures that supply equals demand in each global regions, with primary electrical energy assumed not to be traded (*e.g.*, assumed to be generated and used within a given global region).

Figure 1 gives the structure of the ERB model, as modified for the purposes of the present study. The energy and economic (market-clearing) balances indicated on Fig. 1 are performed for 13 global regions depicted schematically on Fig. 2 (increased from the nine used in the original ERB model[5]) and for nine times separated by 15-year intervals that start in the base year 1975 and moves out to 2095. Energy balance across regions is established by a set of rules[5] for choosing the respective prices that are required for supply to equal demand in each of three energy-service groups for each fuel. The specific test of convergence requires that the difference in regional sums of demand and supply for each of the three fossil primary fuels (oil, gas, and solids) be less than a specified value.

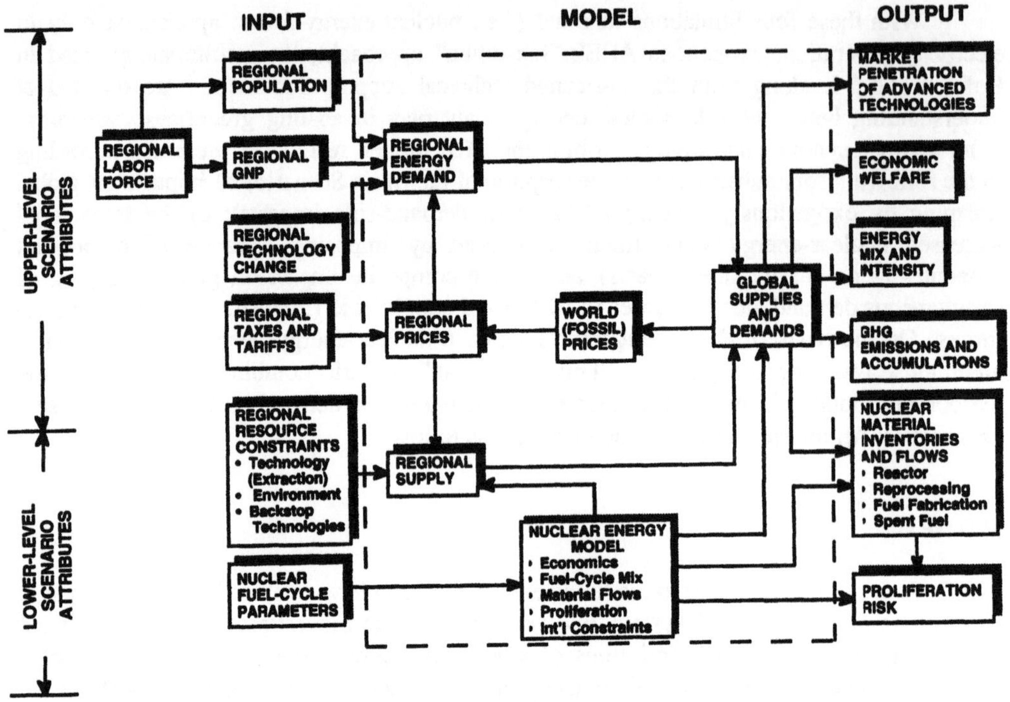

Figure 1 Structural layout of ERB global E^3 model[5] as adapted and modified for the present study. Four main components comprise the ERB economic-equilibrium model: energy demand; energy supply; energy balance; and greenhouse-gas (GHG) emissions. The relationships between inputs and interated outputs, as well as the addition of a (higher fidelity) nuclear energy model (resources, costs, nuclear-materials flows and inventories, and proliferation risk) are also shown.

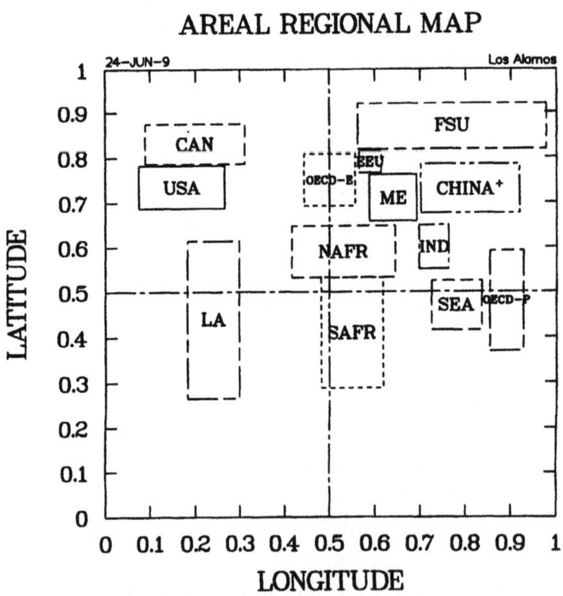

Figure 2 Schematic map of thirteen-region ERB model, with area of each stylized rectangular region reflected the respective land masses. The following regional identifiers are used: 1) USA = United States of America; 2) CAN = Canada; 3) OECD-E = OECD-Europe; 4) OECD-P = OECD-Pacific; 5) EEU = Eastern Europe; 6) FSU = Former Soviet Union; 7) CHINA$^+$ = China plus centrally planned neighboring countries; 8) ME = Middle East; 9) NAFR = North Africa; 10) SAFR = Southern Africa; 11) LA = Latin America; 12) IND = India; and 13) SEA = South and East Asia.

The demand for energy is determined separately for each of the above-mentioned six primary fuels, for each of 13 global regions, and for each of nine times. Five exogenous inputs (including taxes and tariffs) determine the local energy demand. The base (exogenous) GNP (labor-force productivity × population) is used as an indicator of both (regional) economic activity and as an index of regional income. The base GNP is modified through price elasticities to model energy-economy interactions, with GNP $\propto$ price for energy-rich regions and GNP $\propto$ 1/price for global regions that must import energy. More specifically, the demand for energy services (*e.g.*, residential/commercial, industrial, and transportation) for each of thirteen global regions is determined in ERB by: a) the cost of providing these services; b) the level of income (~GNP); and c) the regional population. Energy services are provided by an array of four secondary fuels (liquids, gases, solids, and electricity). The mix of these secondary fuels used to provide a given energy service is determined by a cost-based market-share algorithm, as is the demand for fuels used to produce electricity and the share of oil and gas transformed from coal and biomass. The four secondary energy sources are generated from the six primary fuels, with nuclear, hydroelectric, and solar providing only electrical secondary energy; non-electric solar is treated in ERB as a conservation technology to reduce the demand for the three marketed fuels (*e.g.*, oil, gas, and solids). Modeling of the PE $\rightarrow$ SE $\rightarrow$ ES transformation uses a Leontief-type formulation.[32] As is elaborated in Ref. 6, the nuclear energy module added to ERB, for purposes of the present study, replaces the Leontief equation for nuclear, which originally[5] was based only on a scaled cost of uranium extraction (treated in ERB in this regard like a fossil fuel), with one based on capital, operating and maintenance (O&M), and decontamination and decommissioning (D&D) costs. The resulting nuclear energy cost is then fed back to the ERB demand module to determine the respective market-share fraction as a function of time and region. The uranium resource model originally used in ERB,[5] for purposes of the present study, has been replaced with that suggested in Ref. 33, as interpreted in Ref. 34.

Non-price-induced improvements in end-use energy efficiency are expressed as a time-dependent index of energy productivity that is independent of energy prices and real income. This parameter is similar to the Autonomous Energy Efficiency Improvement (AEEI) used in other more elaborate (inter-temporal) "top-down" models.[21] This approach allows scenarios to be examined that span the range from continued improvement to technological stagnation, irrespective of world energy prices and real income; the limitations of this approach are discussed in Refs. 23 and 35. World energy prices for all fossil fuels are established through energy balance, with regional (fossil) fuel prices being determined by local taxes, tariffs, and transport charges. Interregional trade, however, does not occur for solar, nuclear, or hydroelectric power. In modeling the GHG-mitigating potential of nuclear energy, the AEEI-like parameter ε_k is varied to express the impacts of demand-side solutions, and carbon taxes are applied as a means to allow NC energy sources to assume a larger market share and to reflect supply-side approaches to abating global warming.

While the AEEI parameter is used to examine demand-side impacts, imposition of carbon taxes (C-TAX) at a linear rate ($/tonne/15 yr) starts in the year 2005 (first "future" after the first times 1975 and 1990) to exaime supply-side impacts. Within the context of the version of the ERB model used, these taxes increase the price of fossil-fuel-based energy sources (in proportion to the respective carbon coefficient, kgC/GJ), diminish demand for these energy sources, and diminish economic productivity according to the ERB algorithms [(e.g., GNP $\propto 1/(\text{price})^{\alpha}$]. The carbon taxes *per se* are "removed" from the respective economies, and in this respect the related impact on primarily-energy demand [$\propto (\text{GNP})^{\beta}$] and associated decrease in GHG emission are overestimated; inclusion of both endogenous impacts on AEEI and C-TAX economy coupling/feedback represent important areas of future work.

RESULTS

A range of long-term energy scenarios have been generated based on varying an ensemble of scenario attributes.[9] Nine of the key scenario attributes varied in the Ref.-9 study are summarize on Table I, along with respective ranges of variations. The Ref.-9 study adopted a point-of-departure "Basis Scenario" to which these attribute variations were referenced. The nuclear-energy part of that Basis Scenario invoked a once-through LWR operation, a uranium resource and cost scaling described by a Known Resource (KR) category,[33] and a breeder reactor capital cost that is 50% more than that for LWRs. Without a strong carbon tax to stimulate increased demand for nuclear energy (as well as other reduced- or non-carbon energy sources), these conditions were sufficient to push the economic introduction of breeder reactors to beyond the time frame of this computation (~2100).[8,9,24] For the purposes of the present investigation of the role that nuclear energy might play in reducing the emission of GHGs, the Ref.-9 Basis Scenario with plutonium (mixed plutonium-uranium oxide, MOX) recycle in LWRs is adopted as the Basis Scenario. As for the Ref.-9 study, other (non-nuclear) attributes remain as given in the original version of the ERB model. In this study, implementation of carbon taxes is adopted as the main market force for increasing NC energy supplies while mitigating GHG emission. On the demand side of the equation, increased AEEI is used as a means to examine the relative effectiveness of non-price drivers in reducing GHG emissions. These two supply-side *versus* demand-side approaches to reducing greenhouse warming are then compared. From this comparison, a "nuclear energy scenario" for reduced greenhouse warming is suggested.

Table I. Primary Variables Examined in Ref-9 Study[8,24]

Variable/Attribute	Basis-Case Values
Population growth	UN projections (Fig. 3)[(a)]
Work-force productivity (1/yr)	(0.017,0.025) → (0.009,0.022)[(a)]
AEEI parameter, ε_k(1/yr)	0.0100[(b)]
Carbon tax rate ($/tonneC/15yr)	0 [(c)]
Uranium resource category	CR, KR, or TR[(d)]
NE capital cost, UTC_{LWR}($/We)	2.4 (2095 asymptote)[(e)]
MOX core volume fraction, f^{*}_{MOX}	0.0[(f)]
LMR introduction constraints	(g)
FSB introduction constraints	(h)

(a) 1975 → 2095, dependent on region; varied ± 17% relative to basis case.

(b) reduction in secondary energy required to provide given energy service, independent of region and time (except where noted); varied over range (0.0,0.015) 1/yr.

(c) starting from 2005, rate varied over the range (0,50) $/tonneC/15 yr.

(d) CR = Conventional Resources; KR = Known Resources; TR = Total Resources, per Ref.-33 categorizations; varied over full range; KR adopted as base-case category.

(e) varied over range (2.0,3.0) $/We.

(f) final achieved value, starting in 1995 and increased according to time constant T_{MOX}; independent of region; for T_{MOX} = 10 yr, varied over the range (0.0,0.3).

(g) primarily UTC_{LMR}/UTC_{LWR} = 1.5, KR uranium resources, and no carbon tax; for these base-case conditions, LMRs were too costly for introduction.

(h) primarily minimum conversion ratio allowed and the ratio of UTC_{LMR}/UTC_{LWR} = 1.5 for IFR/LMR fast spectrum burner; the ratio UTC_{LMR}/UTC_{LWR} was varied over the range (1.2,1.5).

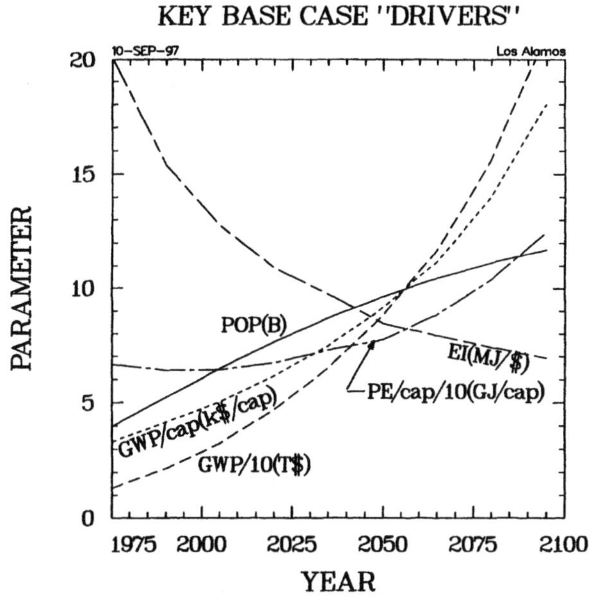

Figure 3 Key base-case aggregated drivers and related first responses: time dependencies of populations, POP (exogenous); GWP (price-adjusted basis case); GWP/POP (price-adjusted basis case); primary-energy intensity, EI = PE/GWP(endogenous), and *per-capita* primary energy, PE/POP(endogenous).

Basis Scenario

The Basis Scenario adopted for this study is largely that used in Ref. 9, but with plutonium recycle in $f_{MOX_f} = 0.3$ of the LWR reactor core volume. Figures 3-9 describe the essential elements of this MOX/LWR Basis Scenario. As for a majority of results presented herein, these results are an aggregate of the 13 global regions described by ERB. The population and GWP drivers behind this scenario are given in Fig. 3, which also includes *per-capita* GWP, *per-capita* primary energy (PE) demand, and the evolution of the global energy intensity, EI(MJ/$) = PE/GWP. Population growth is exogenously inputed from U.N. projections, whereas the exogenous base GWP input is modified to reflect changes in the prices of fossil fuels. The dependence of EI results from endogenous shifts in PE and (to a lesser extent) GWP. The evolving mix of primary energy demand for the Basis Scenario is given on Fig. 4; the diminishing market shares for oil and gas, and the increasing market share for solids (mainly coal, and some biomass) reflect the resource structure (*e.g.*, amount at a given grade and cost) used in the ERB model.[5,17] A moderately disaggregated view of primary- and nuclear-energy demand is depicted in Figs. 5 and 6, respectively. In terms of primary energy, the developing regions (DEV) become comparable users to OECD countries by ~2025, with a similar condition being reached for nuclear energy by 2050. The strong growth in primary- and nuclear-energy growth for the developing countries after ~2050 is driven largely by the CHINA+ region, as is explicitly shown for nuclear energy in Fig. 6B.

The rate of CO_2 emission, R_{CO_2}(GtonneC/yr), for this no-carbon-tax Basis Scenario is shown on Fig. 7. The impact of this carbon-dioxide emission rate on integrated emissions, W_o(GtonneC), accumulated atmospheric CO_2 (carbon), W(GtonneC), and average global temperature rise, ΔT(K), is also shown on Fig. 7. The integrated emissions, W_o, is referenced to atmospheric CO_2 inventories since the dawn of the industrial revolution, which is taken as[15] $t_{IRV} = 1800$, when the atmospheric CO_2 inventory was $W_{IRV} = 594$ GtonneC (2.13 GtonneC/ppmv). The relatively slow increase of the ratio $\Delta T/(W/W_{IRV})$, as determined from the linear integral-response model used,[15] is also included on Fig. 7. Carbon dioxide emissions from each of the 13 regions tracked by the ERB model is shown

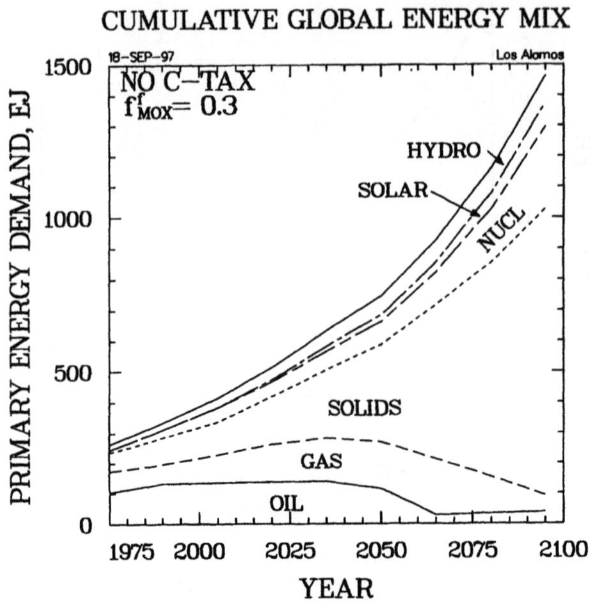

Figure 4 Cumulative evolution of global primary energy mix for the Basis Scenario (solids = coal + biomass): Cumulative primary energy.

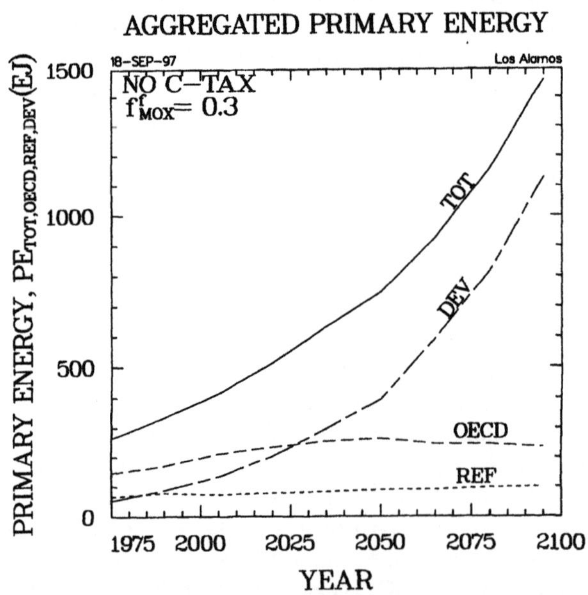

Figure 5 Evolution of aggregated total primary energy for Basis Scenario: OECD = US + CAN + OECD-E + OECD-P; REF + FSU + EEU; and DEV = CHINA^{+} + ME + NAFR + SAFR + LA + IND + SEA.

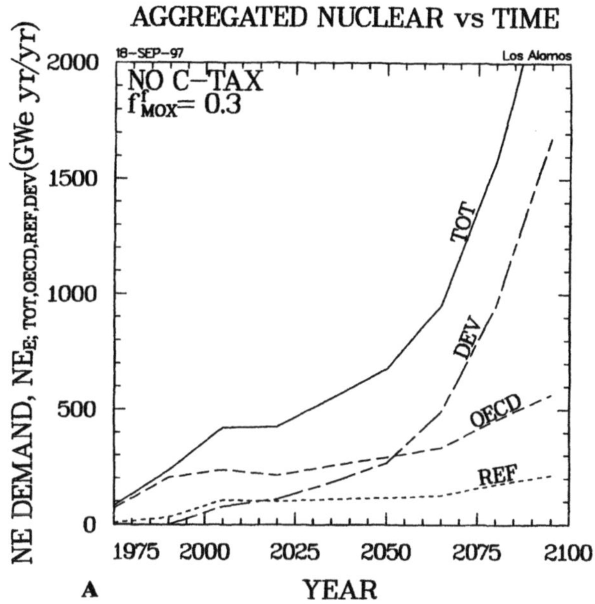

A

Figure 6A Aggregated and (13) regional nuclear energy demand for the Basis Scenario: Aggregated total and macro-regional nuclear energy demand.

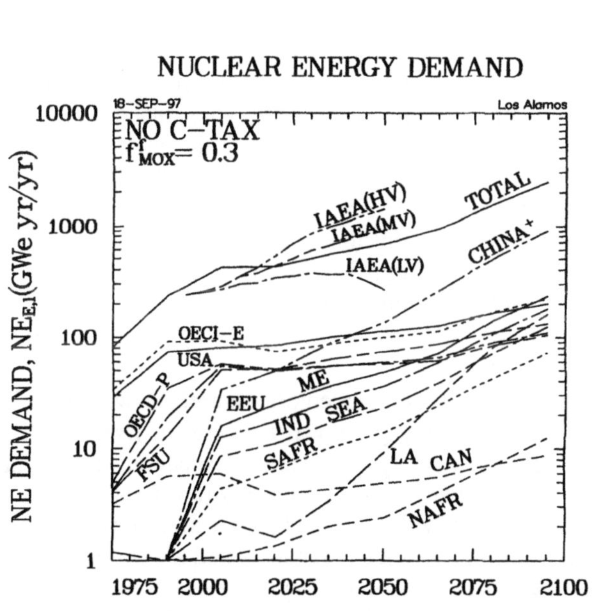

B

Figure 6B Aggregated and (13) regional nuclear energy demand for the Basis Scenario: Regional nuclear energy demand.

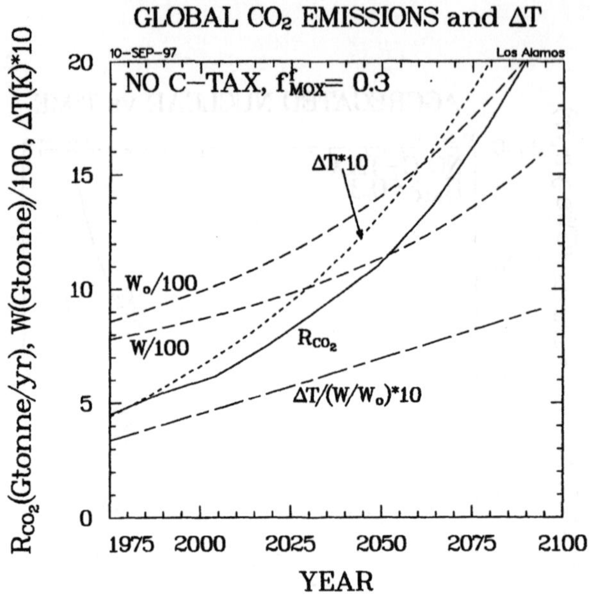

Figure 7 Time dependence of total CO_2 (carbon) emission, integrated emissions, atmospheric accumulation of emissions, and corresponding global average temperature rise, as determined from the linear integral-response model;[15,16] results applied to the zero carbon-tax basis case.[24]

in Fig. 8; the CHINA$^+$ region becomes the dominant contributor of GHGs by the year 2025 for this Basis Scenario.

The evolution of plutonium inventories by region is used in a multi-attribute utility (MAU) analysis,[10] that has been synthesized from earlier work[11-14] for use in the ERB nuclear model, to yield relative measures of a utility for proliferation, <u>, and a proliferation risk index, PRI. These relative (and highly subjective) measures of proliferation risk are adopted as the primary non-economic "cost" for nuclear energy against which any benefit of reduced GHG emission is measured. The PRI is a time-discounted sum of regionally weighted utilities evaluated from the viewpoint of a proliferator, which in turn is a weighted average of subutility functions that in turn reflect proliferator-oriented attributes that measure cost, technological difficulty, detection risk, and material availability;[10] the PRI indicates a value-based potential for proliferation rather than a probability for proliferation, is measured on a scale from zero to unity, and is subjectively evaluated from the perspective of a given region (*e.g.*, the U.S.).

The buildup of global plutonium inventories correlates with the relative CO_2 (carbon) accumulation, W/W_{IRV}, or the average global temperature rise, $\Delta T(K)$, that results. The latter is computed from the year t_{IRV}. These correlations are central to subsequent correlations of global climate change (GCC), nuclear-proliferation, and economic impacts. Specifically, the risks associated with increased global inventories of plutonium and GHGs are expressed in terms of the PRI and ΔT parameters and are correlated in terms of a reduced "operating curve" for the Basis Scenario on Fig. 9. As important as is the need to translate both PRI and ΔT into economic and social consequences, the present study does not advance beyond the correlation shown given in Fig. 9. This "operating curve" *per se* is not as important to understanding proliferation-risk/GCC/GWP connectivities as are relative shifts in the slopes and magnitudes at a given time as key scenario drivers (*e.g.*, carbon tax rates or exogenously driven AEEI trajectories) are changed. Figure 9 also compares the PRI impacts (for the no-carbon-tax case) of plutonium recycle (f_{MOX_f} = 0.30) and the use of the once-

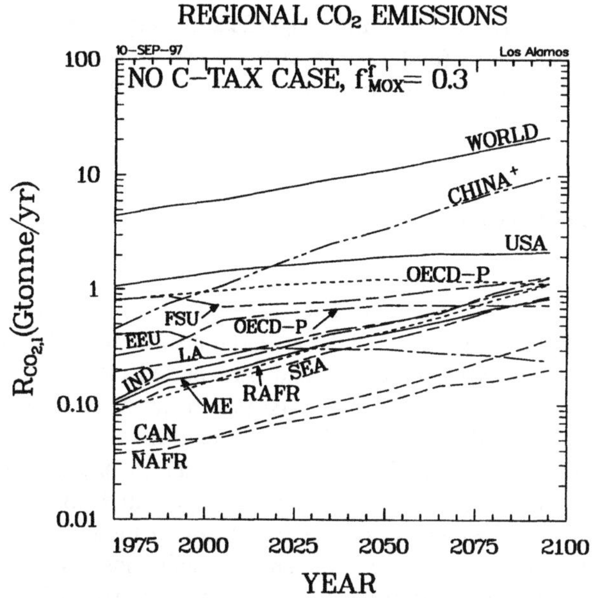

Figure 8 Atmospheric carbon emission rates as a function of time and region for Basis Scenario.

through (LWR) fuel cycle. Plutonium recycle increases the PRI by ~10% while having little impact on reducing GCC impacts (*e.g.*, ΔT). Actually, the lines of constant time (an isochrone for 2095 is shown on Fig. 9) are almost vertical, with a slight off-vertical orientation indicating that the small added cost associated with the f_{MOX_f} = 0.0 → 0.3 transition, which increases the cost of nuclear energy slightly and reduces (slightly) demand, results in a small increase in fossil-fuel use, and leads to somewhat larger values of ΔT

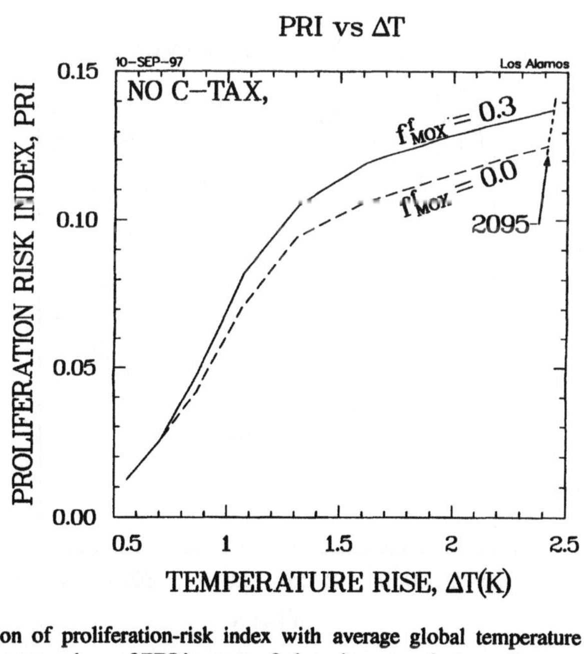

Figure 9 Correlation of proliferation-risk index with average global temperature rise for case without carbon tax imposed; a comparison of PRI impacts of plutonium recycle (*e.g.*, f_{MOX_f} = 0.0 *versus* 0.30) is shown.

(< 0.05 K) for the f_{MOX_f} = 0.30 case. Significantly larger impacts are computed for enhanced use of nuclear energy (and other reduced- or non-carbon energy sources) forced by imposing carbon taxes, however. Before results of implementing this supply-side driver are reported, however, the impact of variations in the demand-side parameter AEEI are first reported.

Demand-Side Impacts: AEEI

The parameter ε_{jk}(1/yr) represents a non-price-induced reduction in the amount of secondary energy j (j = liquids, gases, solids, and electricity)[5] needed to provide an energy service k (k = residential/commercial, industrial, and transportation). For the Basic Scenario, ε_k (the j subscript is not used) after the second time period (1990) is 0.0100 1/yr for all regions and all times. As noted in Ref. 35, the parameter AEEI is not well named; as a measure of non-price induced changes in EI, it may neither be autonomous nor deal solely with energy efficiency. The AEEI parameter attempts to account for the impacts of technological developments, (economy) structural changes, and policy-driven inducements in the move towards increased energy efficiency. Many of these forces reflected in AEEI-like effects are endogenous to the economic-energy evolutionary process, and cannot be specified as an external driver. Reference 22, in fact, reported AEEI-like effects from a sectoral model of the economy without explicitly introducing the AEEI parameter.

The scenarios considered under "AEEI variations" (Table I) examine impacts over the range ε_k = (0.0,0.015), where again ε_k is regionally and temporally (after 1990) constant at the designated value. One case, ε_k = 0.015(RAMP), linearly ramps ε_k down from 0.015 (in 1990) to 0.0 in 2095. These impacts are summarized on Figs. 10-13. Specifically, the impact on primary- and nuclear-energy demand is depicted on Figs. 10 and 11. The reflection of these changes in end-use efficiency on the energy intensity (again, starting in 1990) indicates that ε_k values much below ~0.0050 1/yr, in a globally aggregated sense, freeze any improvement (*e.g.*, decrease) in the global energy intensity, EI(MJ/$).

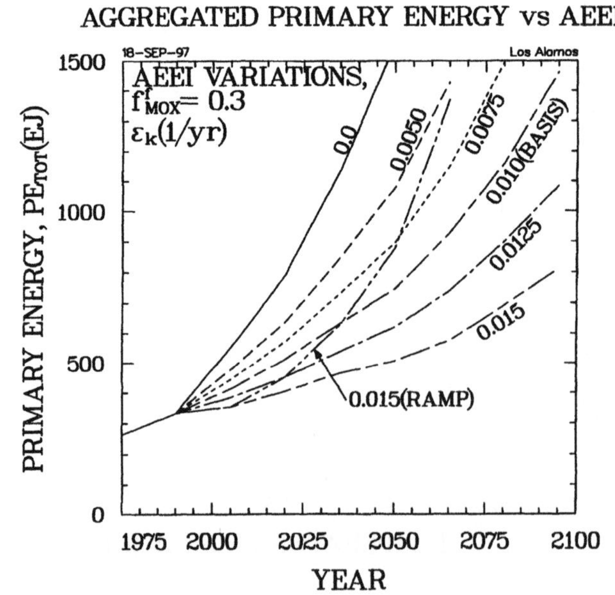

Figure 10 Primary energy demand as a function of time and AEEI; the ε_k = 0.015(RAMP) case starts ramping from the indicated value in 2005 and linearly decreases to zero by 2095.

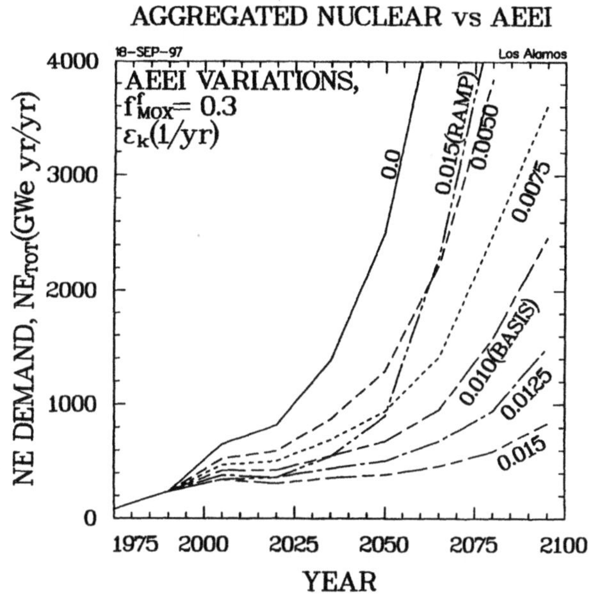

Figure 11 Nuclear energy demand as a function of time and AEEI; the ε_k = 0.015(RAMP) case starts ramping from the indicated value in 2005 and linearly decreases to zero by 2095.

The range of ε_k values considered not only has a significant impact on primary-energy demand (Fig. 10), but relatedly leads to wide swings in carbon-dioxide emissions, as is shown on Fig. 12. The average global temperature rises that result are depicted on Fig. 13. That decreases in ε_k below the 0.0100 1/yr basis-scenario value make an already serious problem more serious comes as no surprise; that 50% increases in ε_k have such

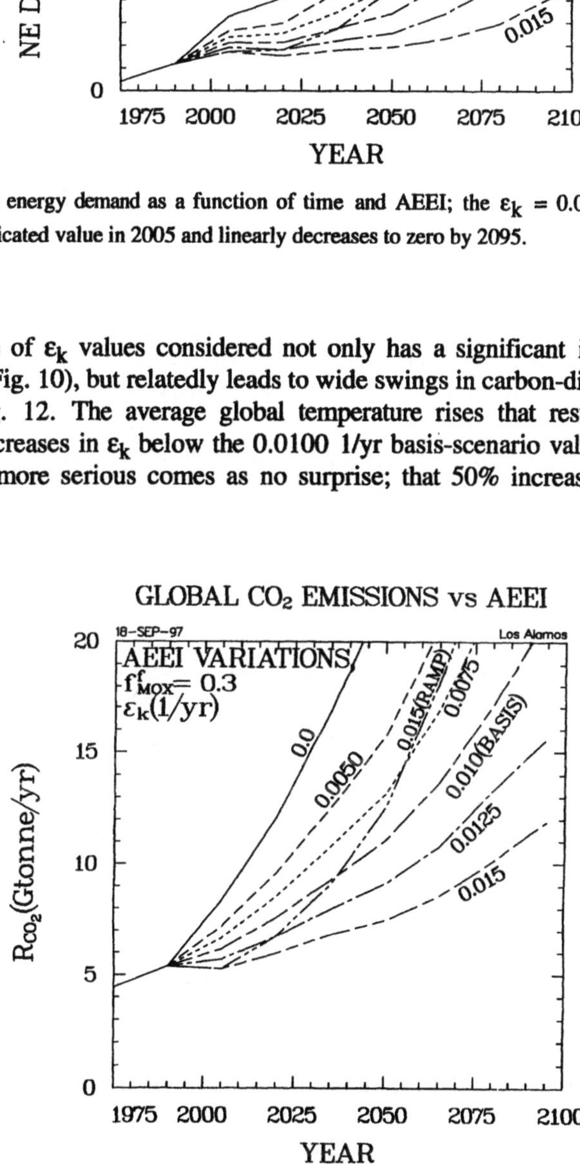

Figure 12 Carbon-dioxide (carbon) emission rate as a function of time and AEEI; the ε_k = 0.015(RAMP) case starts ramping from the indicated value in 2005 and linearly decreases to zero by 2095.

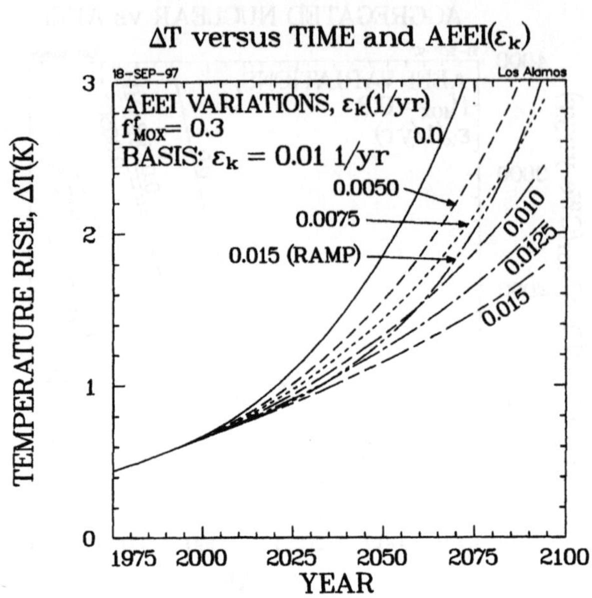

Figure 13 Average global temperature rise as a function of time and AEEI; the ε_k = 0.015(RAMP) case starts ramping from the indicated value in 2005 and linearly decreases to zero by 2095.

relatively weak impact on mitigating global warming is. Essentially, across-the-board increases in AEEI result in needed, but insufficient, decreases in GHG emissions; this parameter alone cannot induce changes in the primary-energy mix needed to move aggressively to NC energy sources. The implementation of the supply-side forces embodied in energy taxes based on carbon content can cause such a shift. Unfortunately, if applied regressively, the increased prices that result can drive decreased productivity. These issues are examined, within the limitations of the ERB model, in the following section.

Supply-side Impacts: Carbon Taxes

Energy Demand and Mix. A carbon tax is applied to fossil fuels in proportion to carbon content per unit of released energy. Beginning in 2005, this carbon tax is applied for linearly increasing rates ranging from 0 to 50 \$/tonneC/15yr; hence, for a rate of 40 \$/tonneC/15yr, the carbon tax at the last time considered (2095) would be 240 \$/tonneC. This carbon tax schedule puts the heaviest penalty on coal (23.8 kgC/GJ) and the least penalty on natural gas (13.7 kgC/GJ), with oil being intermediate (19.2 kgC/GJ). According to the ERB algorithms, carbon taxes increase the price of the affected energy source, decrease the market share for that energy source, and reduce the price-based adjustments to the (exogenous) base GNPs. The relationship between energy prices and GNP used in the ERB model derive from the oil shocks of the 1970s, and, as a result, the GNP losses reported by ERB "are unreliable and excessive."[35] In spite of a warning against use of the GNP figures generated by ERB, GNP decrements, ΔGNP, are reported here, along with total cost (tax) figures.

In its present form, collected carbon taxes are not returned to the GNP, but are simply allowed to "disappear". An algorithm was added to ERB to monitor both actual and present-valued carbon taxes and GNP decrements related thereto; these are reported here as a first step towards developing a more sophisticated (*e.g.*, revenue-neutral, import tariffs based on carbon content, *etc.*) carbon tax schedule. For the purposes of the present study, the imposition of carbon taxes is used primarily as a means to increase the price of fossil fuels and to increase the market share of NC energy sources.

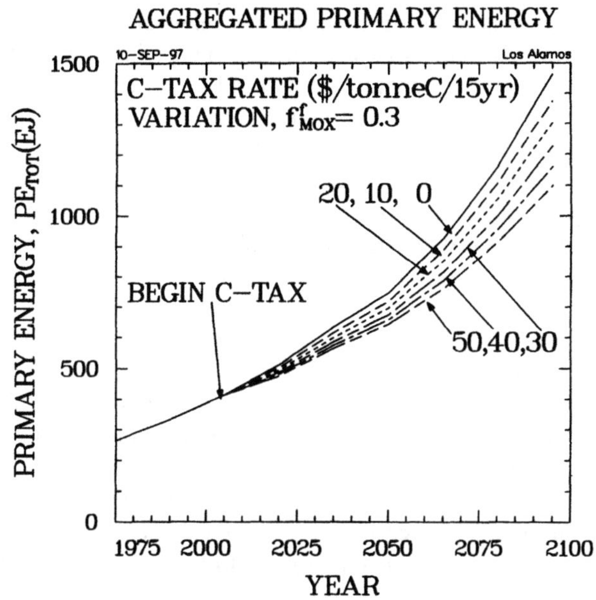

Figure 14 Primary energy demand as a function of time and carbon tax rate, starting in 2005.

The impact of carbon taxes on primary energy use is shown on Fig. 14; at the highest rate of carbon taxation, primary energy use in 2095 could be reduced by ~25% relative to the Basis Scenario. The shift in market shares for the six primary energy sources from the Basis Scenario (no carbon tax) to the case of maximum carbon tax rate (50 $/tonneC/15yr) is as follows; coal looses the greatest market share (~65% → 22% in 2095), nuclear and solar (electric) energies show a strong increase in market share (~19% → 46% and ~5-6% → 13% in 2095, respectively), resource-limited hydroelectric shows only a moderate increase, and

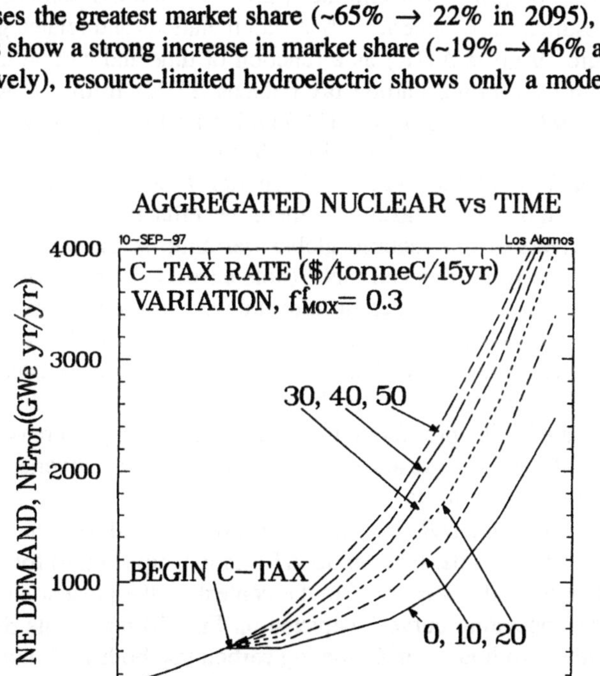

Figure 15 Nuclear energy demand as a function of time and carbon tax rate, starting in 2005.

gas, while diminishing somewhat in time, shows relatively little change from the Basis Scenario. The shift towards more solar and nuclear energy infers an increase in the use of electricity; the fraction of primary energy that is used to generate electricity increases from ~16% to 22% in 2095 for the maximum carbon tax rate.

Focusing on nuclear energy, Fig. 15 gives the dependence of annual nuclear energy demand on carbon tax rate. For the 50 $/tonneC/15yr carbon tax rate, nuclear energy demand increases in 2095 by ~43% relative to the basis scenario. The required deployment rate for this case increases from ~75 GWe/yr to ~85 GWe/yr (for an 80% plant availability factor). Similar deployment rates are required in the out years for the no-carbon-tax case. Figure 16 gives the (same basis) fraction of primary energy demand satisfied by nuclear energy, which in the out years increases from ~18% for the Basis Scenario to ~45% for the most agressive carbon tax rate considered.

Under these circumstances, nuclear energy becomes a major player in the world energy mix. The reduction in atmospheric CO_2 (carbon) emissions that accompanies this carbon-tax-induced increase in nuclear (and solar) energy demand is illustrated in Fig. 17, which also gives *per-capita* and per-primary-energy carbon emission for the Basis Scenario. For the latter, while carbon release per unit of primary energy decreases somewhat, more of this reduced-carbon energy is being used on a *per-capita* basis as prosperity drives a global *per-capita* appetite for energy.

Figure 18 gives a composite curve of fractional reduction of CO_2 emissions ($\Delta R_C/R_C$, relative to the zero-carbon-tax Basis Scenario) as a function of the carbon tax, UC_{TX}($/tonneC), as assembled from the five carbon tax rates being considered. Shown also on this figure is the result of a regression fit to seven econometric, optimization, and hybrid models, as is reported in Ref. 38.

Using the integral-response formulation reported in Ref. 15, and adopting t_{IRV} = 1800 as the beginning of the industrial revolution and the beginning of anthroprogenically driven global warming (W_{IRV} = 594 GtonneC, $\Delta T = 0$), the CO_2 emission rates given on Fig. 17 are used to estimate atmospheric carbon accumulations and related global temperature rises, $\Delta T(K)$. Figure 19 gives $\Delta T(K)$ as a function of time and carbon tax rate. In the out years, the application to a strong carbon tax reduces $\Delta T(K)$ from 2.4 K to 1.4 K; these temperature rises are referenced to t_{IRV} = 1800 and, based on the model used, has already reached ~0.4 K by the start of this computation (1975).

E^3-Trilemma Trade Offs. Whatever "benefits" accrue from the mitigation of global temperature rise (through carbon taxation), these benefits must be compared to "costs" associated with the drivers of this reduced global warming. In the present context, some of these costs are economic [*e.g.*, reduced GNP (note caveats given previously[35]) and an (as yet) unallocated tax stream], some are environmental (*e.g.*, waste streams associated with the increased use of NC energy sources, which are primarily solar and nuclear), and some are social-political (*e.g.*, increased risks and altered social structures associated with the need to reduce risks related to nuclear-weapons proliferation). The following discussions deal first with trade off associated with proliferation risk that accompany increased use of nuclear energy, and then is followed by a discussion of some aspects of adverse economic impacts of imposing unallocated carbon taxes.

Proliferation versus GCC Risks. At the level of this analysis, the culmination of the comparative risk assessment is the PRI *versus* ΔT relationship (Fig. 9) for this special set of carbon-tax-driven (supply-side) scenarios. In the context of the present study, the evolution of the PRI *versus* ΔT "operating curves" depicted on Fig. 20 represents the final result. As discussed above, with or without a GHC-abating carbon tax, both PRI and ΔT will increase with time as populations in number and living standard develop. The first frame of Fig. 20 gives this PRI *versus* ΔT evolution with increasing carbon tax rates, whereas the second frame stresses more the increased nuclear-energy share under imposition of carbon taxation by giving the fractional increase in PRI relative to the zero-tax case as a function of ΔT. The

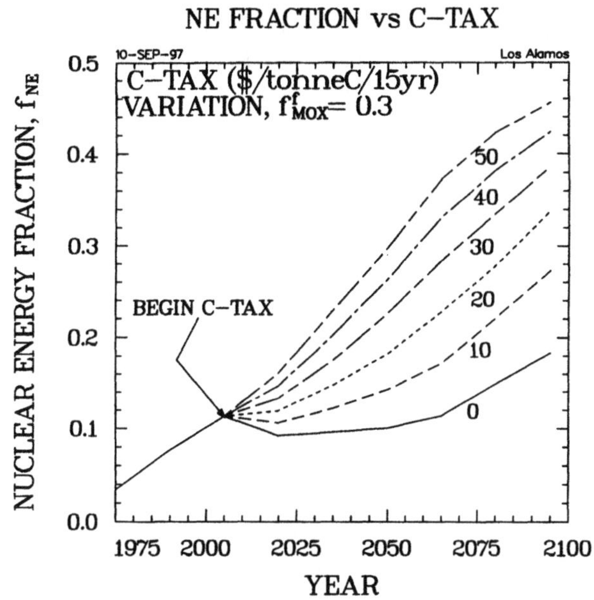

Figure 16 Nuclear energy as a fraction of total primary energy as a function of time and carbon tax rate, starting in 2005.

added sensitivity of plotting $\Delta PRI/PRI_0$ reveals that, for a given taxation rate ($/tonneC/15yr), the fractional increase in PRI shows a maximum at ~2065 that is independent of the rate at which the carbon tax is applied. Generally, increased use of nuclear energy through the imposition of a carbon tax slows the rate of global warming while increasing proliferation risk a few percent relative to the zero-carbon-tax Basis Scenario.

Economic versus GCC Risks. Resolution of the economic costs of this particular set of drivers, as monitored through GNP impacts and unallocated carbon taxes, remains as

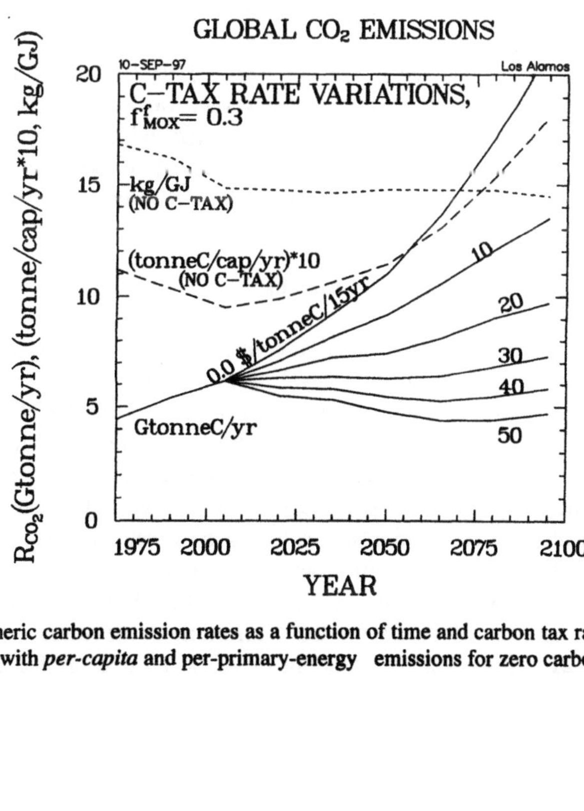

Figure 17 Atmospheric carbon emission rates as a function of time and carbon tax rate, beginning in 2005; emission rates along with *per-capita* and per-primary-energy emissions for zero carbon taxes are shown.

future work that must ultimately relate abatement costs to achieve a given reduction in ΔT to damage costs associated with accommodation to GCC impacts; these costs are sometimes expressed as percentages of GNP.[39,40] For present purposes, a unit cost of CO_2 abatement, UC_A(\$/tonneC), is define as the ratio of reduced CO_2 emissions relative to the Basis Scenario, $\Delta R_{CO_2} = R_{CO_2}$(No C-TAX) – R_{CO_2}(C-TAX), to either the total carbon taxes collected for that year, TAX, or the sum TAX + ΔGWP, where ΔGWP = GWP(No C-TAX) – GWP(C-TAX). Figure 21 gives the time dependence of ΔT, TAX, ΔGWP, and these two ways of calculating UC_A. Also shown is the ratio TAX/ΔGWP varying from 2.4 in 2020 to 0.6 in 2095. Attempts to correlate both measures of abatement unit cost with the unit carbon tax, UX_{TX}, are reported on Fig. 22 for the range of carbon tax rates being considered. Based on $UC_{A,TAX} = TAX/\Delta R_{CO_2}$, high tax rates favor lower "abatement costs" by a factor of ~2. On the other hand, for $UC_{A,TAX + \Delta GWP} = (TAX + \Delta GWP)/\Delta R_{CO_2}$, higher carbon taxes result in ~15% higher "abatement costs". If a "revenue-neutral" carbon tax scheme could be devised and implemented, then TAX + ΔGWP could be reduced in magnitude (and possibly sign).

Some would argue that both TAX(t) and ΔGWP(t) should be discounted at a rate DR(1/yr) to a reference year, summed over the computational period, and expressed in present-value form. Figure 23 gives the decrease in world GNP as a function of the rate of carbon taxation. These GWP percentage decreases are expressed in two forms: a) the percent decrease in the last-year (2095) GWP with and without a carbon tax imposed at a given rate; and b) the percent decrease in the present value of all GWPs over the study period, using a constant pure discount rate of DR = 0.04 1/yr; the former gives $(\Delta GWP/GWP)_{2095} = 4\%$, and the latter gives $(\Delta GWP/GWP)_{PV} = \sim 0.7\%$. The ratio of the present value of incremental GWP to the present value of all carbon taxes collected over the computation period, again using DR = 0.04 1/yr, is nominally constant in the range 0.6-0.7; the present value of all carbon taxes collected over the computation period is about twice the present value of the GWP decrement. Again, the previously stated caution about using price-adjusted GNP values from ERB should be kept in mind. Also shown on Fig. 22 is the decrease in atmospheric CO_2 accumulation (again, W_{IRV} = 594 Gtonne is the atmospheric carbon inventory for t_{IRV} = 1800). This reduction in global warming might be considered a benefit against which the decreased GWP could balance, albeit, a more careful and consistent accounting of the collected carbon taxes, as well as a weaker price-GNP scaling,[35] could reduce or reverse the GWP decrements computed from the present model. The percentage increase in proliferation risk evaluated in the last year, $(\Delta PRI/PRI)_{2095}$, associated with the increased implementation of nuclear energy is also shown on Fig. 23. While ΔPRI is small relative to PRI, no quantitative statement can be made with respect to this increased proliferation risk attendant to increased use of nuclear energy to abatement GHG accumulation until the consequences of PRI without carbon taxes are fully assessed.

Composite Demand-Side/Supply-Side Impacts

The relative impacts on stemming greenhouse warming through demand-side (increased AEEI, $\varepsilon_k = 0.0100 \rightarrow 0.0150$ 1/yr), supply-side (carbon tax rates, $0 \rightarrow 50$ \$/tonneC/15yr), and a combination of both are given a cursory examination in this section. Along with the Basis Scenario ($\varepsilon_k = 0.0100$ 1/yr, no carbon tax), the four cases listed in Table II are compared. Figures 24 and 25 give the time dependence of primary-energy and nuclear-energy demand, respectively, for these four case. For a given unit carbon tax, the 25% increase in ε_k results in ~10% additional decrease in the relative CO_2 emission rate. The average global temperature rise for all four cases are summarized on Fig. 26. The bulk of the ~45% decrease in ΔT comes from the supply-side carbon tax, with AEEI contributions being relatively minor. The impact of AEEI on the approximate measures of abatement cost, UC_A(\$/tonneC, Fig. 22), however, can amount to ~33% reductions for the case of UC_A

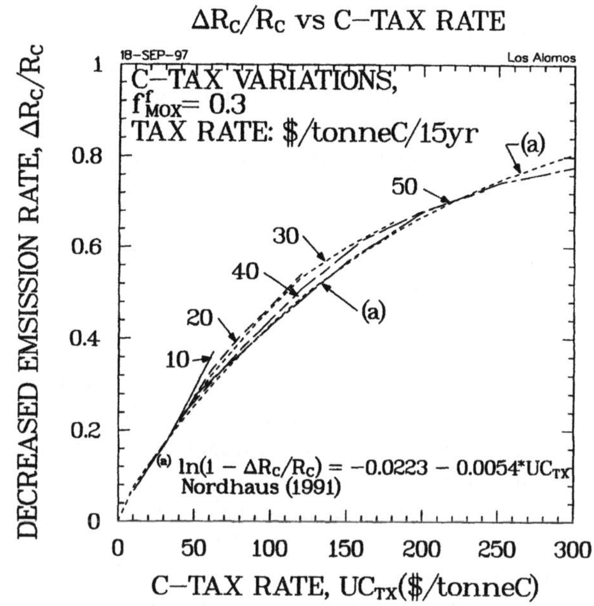

Figure 18 Decrease in atmospheric carbon emission rate as a function of carbon tax for a range of carbon tax rates; a regression fit to a number of model predictions[38] is also shown.

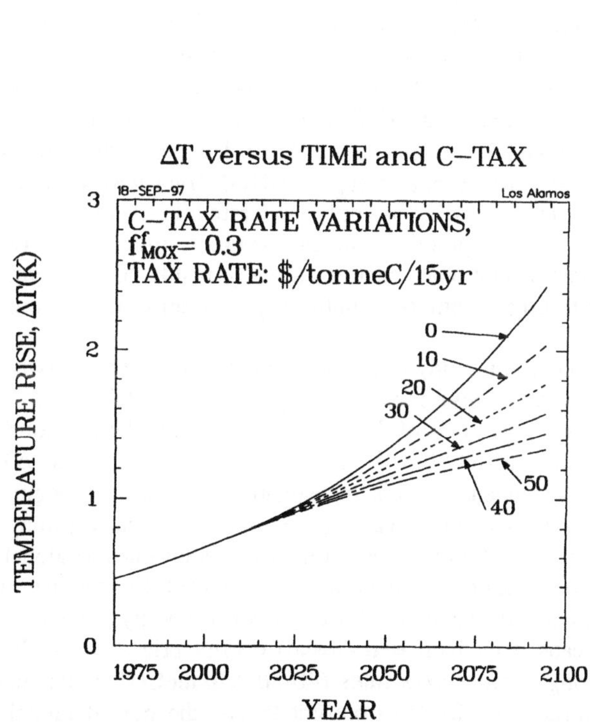

Figure 19 Time dependence of average global temperature rise for a range of carbon tax rates, starting in 2005.

based only on TAX. For the case of UC_A based on TAX + ΔGWP, the cost reduction for superposing the demand-side abatement solution onto the supply-side solution amounts to ~23%. Hence, while the latter has only a minor impact on reducing ΔT *per se*, a strong economic symbiosis in combining the two may exist. Lastly, a direct comparison of increased proliferation risk (PRI) that accompanies the decreased GCC risk (ΔT) is given in Fig. 27; the combined C-TAX + AEEI attack on global warming reduces somewhat PRI relative to a purely supply-side (carbon tax) strategy, while giving an added (slight) reduction in global warming. A central question, however, is the abatement cost associated with demand-side approaches to reducing GHG emissions.[18,40]

SUMMARY AND CONCLUSIONS

A range of long-term futures for nuclear energy have been examined in Ref. 9 by building "surprise-free" scenarios using a consistent, but simplified, modeling tool.[5] Defining scenario attributes are placed in a hierarchy that divides determinants of nuclear energy futures into external forces and forces that originate from within nuclear energy *per se*. By varying the former upper-level scenario attributes (*e.g.*, population, workforce productivity, energy intensity or end-use transformation efficiency, global taxes, top-level nuclear energy economics), a wide range of nuclear energy demand scenarios can be generated. Although these scenarios represent only possibilities, and are not predictions, they nevertheless provide a quantitative basis and connectivity for examining impacts of the lower-level internal drivers that influence directly the economic and operational character of nuclear power. The OT/LWR Basis Scenario adopted in Ref. 9 as a point-of-departure case was modified to include MOX recycle and provided the Basis Scenario for the present study of the impacts of nuclear energy on greenhouse warming. Carbon taxes where used as a supply-side "forcing function" to increase market share of key NC energy sources (mainly solar and nuclear energies). Top-level economic and proliferation-risk implications of this demand-side approach to reducing GHG emissions were examined. As a representative demand-side driver of GHG abatement, the AEEI-like parameter used to define the no-carbon-tax MOX/LWR Basis Scenario (ε_k = 0.0100 1/yr) was varied. It was found that while (exogenously) increased AEEI has only moderate impacts on greenhouse warming *per se* (Table II), when used in conjunction with carbon taxes, important decreases in (the highly approximate) measures of unit abatement costs, UC_A result. Similar symbiotic effects may also come into play through attempts to mitigate proliferation risks along with GCC risks (Fig. 27).

A central theme of this study is embodied in the relationship between economic (*e.g.*, ΔGWP, TAX, UC_A, *etc.*), environmental (*e.g.*, GCC, proliferation), and energy (*e.g.*, AEEI, PE mixes, EI, *etc.*) elements of the E^3 equation. While the relationships demonstrated quantitatively herein are generally based on relative metrics (*e.g.*, PRI, ΔT, UC_A, *etc.*) and are far from being comprehensive, this investigation represents a beginning. Specifically, using the proliferation risk index (PRI) and the estimate of global warming generated from a linear integral-response model[15] that relates GHG emission rates to global temperature rise, ΔT, a range of carbon-tax-driven scenarios was created to examine tradeoffs between increased PRI that accompanies increased use of nuclear energy, decreased global warming, and reduced GWP caused by increased (fossil) energy prices (Figs. 20 and 21). It was found that while strong carbon taxes rates (40-50 \$/tonneC/15yr, beginning in 2005) can return CO_2 emission rates in ~2100 to present levels, the rate of global temperature rise, while significantly diminished, remains positive (~0.5 K/yr, compared to 2.8 K/100yr for the case of no carbon taxes). In the 2100 time frame, GWP would be reduced by 3-4% (~0.8% on an integrated present-value basis using a 4%/yr pure discount rate), primary energy used would be reduced by 20-25%, and nuclear energy would experience a ~80%

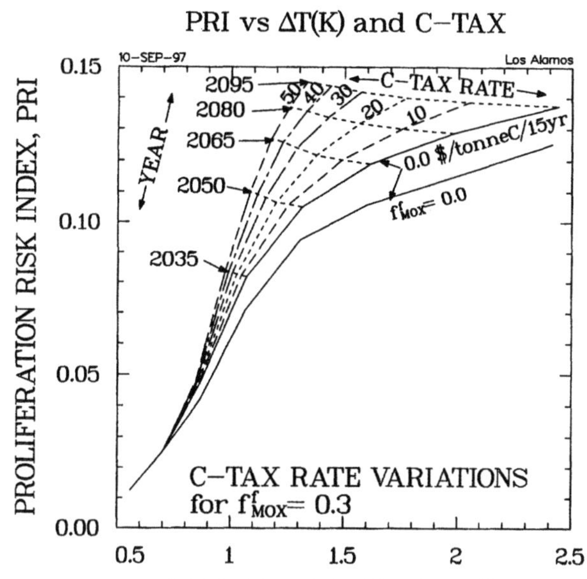

A

Figure 20A Direct comparison of proliferation-risk-index *versus* atmospheric temperature-rise "operating curves" as the rate of carbon taxation is varied: direct comparison of PRI *versus* ΔT, showing isochrones.

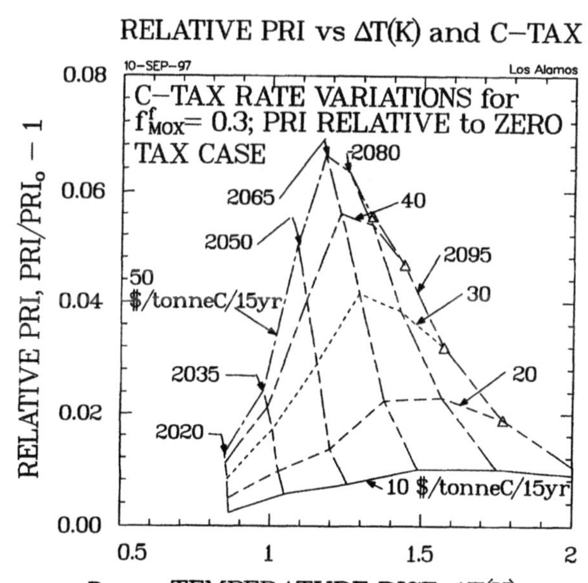

B

Figure 20B Direct comparison of proliferation-risk-index *versus* atmospheric temperature-rise "operating curves" as the rate of carbon taxation is varied: change in PRI relative to the no-carbon-tax case, showing isochrones.

increase (necessitating a deployment rate of ~80 GWe/yr in the out years around 2100). The ratio of present value of all carbon taxes to present value of lost GWP (again, using a 4%/yr pure discount rate) amounts to ~1.3 over most of the computational period. The PRI is increased by only 5-6% for the maximum nuclear-energy implementation (*e.g.*, strongest carbon tax rate) in ~2100. Specifically, the explicit relationship between these relative

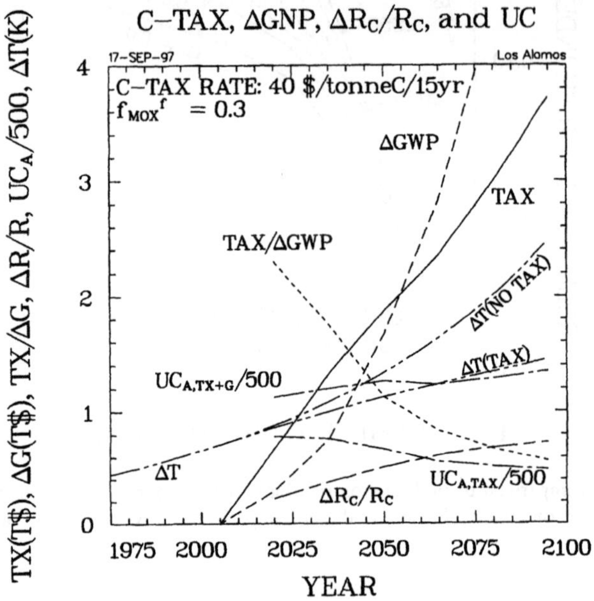

Figure 21 Time dependence of total carbon taxes, decreased GWP, tax-to-GWP ratio, percent decrease in atmospheric carbon emission rate, unit cost of CO_2 abatement, and average global temperature rise for a carbon tax rate of 40 $/tonneC/15yr.

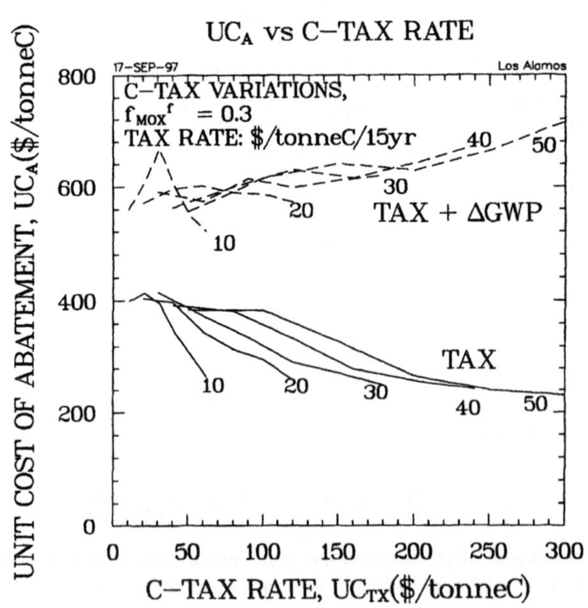

Figure 22 Dependencies of two measures of unit cost of abatement, UC_A($/tonneC), as a function of unit carbon tax, UC_{TX}($/tonneC), for a range of carbon tax rates ($/tonneC/15yr).

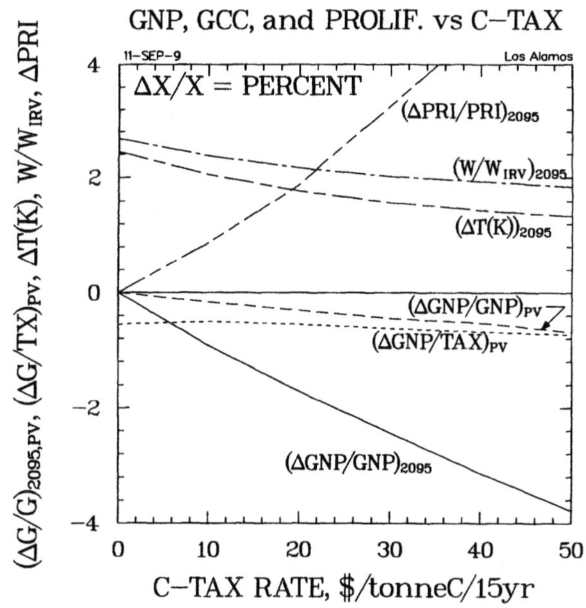

Figure 23 Impact of carbon tax rates on either present value (PV) or last-year (2095) gross productivity (G = GWP), comparing GCC parameters (W/WIRV and ΔT) with proliferation parameters (PRI); all relative changes ΔX/X are expressed as percentages: direct dependence on carbon tax rate.

measures of (increased) proliferation risk and (decreased) global temperature rise (Fig. 20) indicates that for this 5-6% increase in PRI, ΔT in 2100 is reduced form 2.4 K for the no-carbon-tax case to 1.4-1.5 K, but, again, global temperature continues to rise at a rate of 0.5 K/100yr in 2100 for the strong carbon tax rates. These correlative results between proliferation risk and GCC impacts, however, project only relative trends; the "real" implications of the base (*e.g.*, for no carbon tax) PRI growing to ~0.14, ± 5-6% with or without carbon-tax-induced growths in nuclear energy, along with the assessment of "actual" impacts of decreasing the global temperature rise by ~ 1 K over ~100 years needs resolution.

Table II. Summary of Cases Used to Compare Demand-Side *versus* Supply-Side Approaches to Mitigating Greenhouse Warming

Scenario Identification	Carbon Tax Rate ($/tonneC/15yr)	AEEI ε_k (1/yr)	Percentage Changes in 2095 Values PE	NE	EI	R_{CO_2}	ΔT
Basis	0	0.0100	--	--	--	--	--
Carbon Tax	40	0.0100	−20.	+83.	~0.	−73.	−41.
AEEI	0	0.0125	−25.	−39.	−29.	−29.	-5.
AEEI + C-TAX	40	0.0125	−43.	+22.	−39.	-83.	−45.

AEEI autonomous energy efficiency improvement
ε_k = annual rate of increase of SE → ES conversion efficiency
SE = secondary energy
ES = energy service
PE = primary energy
NE = nuclear energy
EI = primary energy intensity, PE/GWP(MJ/$)
R_{CO_2} (GtonneC/yr) = CO_2 emission rate
ΔT (K) = average global surface temperature rise

Finally, an emissions scenario base case was synthesized from the Basis Scenario by implementing both supply-side (carbon tax rate increased from 0 to 40 \$/tonneC/15yr) and demand-side (AEEI-like parameter ε_k increased from 0.0100 1/yr to 0.0125 1/yr) drivers. As in indicated n Fig. 28, this 25% increase in (global) AEEI reduces the amount of nuclear energy required for a ~90% reduction in global warming (in 2095) by ~30%.

Throughout the narrative, a number of shortcomings and areas of future work were identified, many of which are related to differences in "top-down" *versus* "bottom-up" modeling approaches.[23,35] These shortcomings and areas of future work are summarized as follows:

- *AEEI Parameter:* The simplified, exogenous variation of the parameter ε_k only approximates complex, endogenous interactions between technology improvements, (economy) structural (sectoral) interactions, and policy-driven behavioral changes;[35] schemes to endogenize AEEI-like parameters should be investigated.
- *GNP Feedback:* As approximate as the price-GNP feedback in ERB is, the calibration of this feedback is based on responses to the oil crises of the 1970s and, hence, may be overly responsive;[35] re-calibration and parametric sensitivity studies need to be performed.
- *Carbon Taxation:* The impact of carbon taxes on GNP, as modeled in ERB, is only though the above-described price-GNP feedback; no attempt has been made in these computations in enforce revenue-neutral (or better) schemes to return these taxes to the regional GNP streams; higher-fidelity taxation and (carbon) rights-trading schemes must be investigated.
- *Regional Variations:* The results presented herein pertain to a generally uniform globe; no attempt was made to tailor rates of carbon taxation, AEEI improvements, or nuclear-energy deployment on a regional basis to optimize all elements of the E^3 equation on a global basis; region-based growth scenarios have been shown to be important[17] and must be developed further.

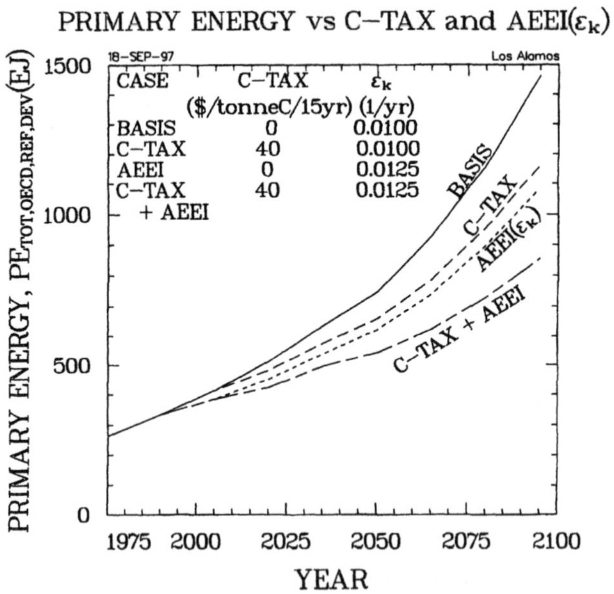

Figure 24 Primary energy demand as a function of time for a combination carbon tax rate and AEEI, showing relative impacts of supply-side (carbon tax) and demand-side (AEEI) scenario attributes.

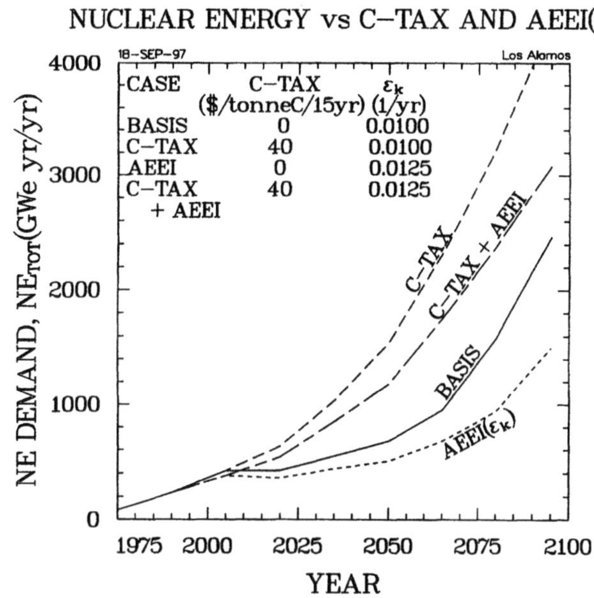

Figure 25 Nuclear energy demand as a function of time for a combination carbon tax rate and AEEI, showing relative impacts of supply-side (carbon tax) and demand-side (AEEI) scenario attributes.

- *Quantitative Metrics:* While the GNP impacts are quantitative, in spite of the limitations listed above, the GCC metric (ΔT) and the proliferation risk metric (PRI) remain qualitative in terms of real economic welfare impacts; attempts must be made to quantify economic impacts of PRI and ΔT.

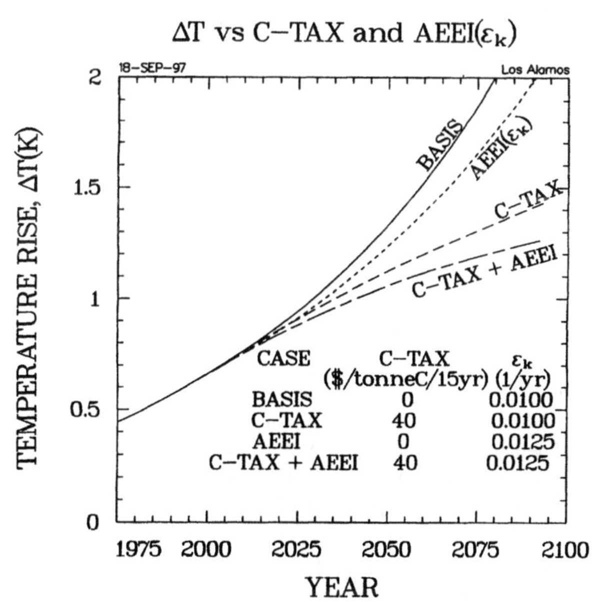

Figure 26 Average global temperature rise as a function of time for a combination carbon tax rate and AEEI, showing relative impacts of supply-side (carbon tax) and demand-side (AEEI) scenario attributes.

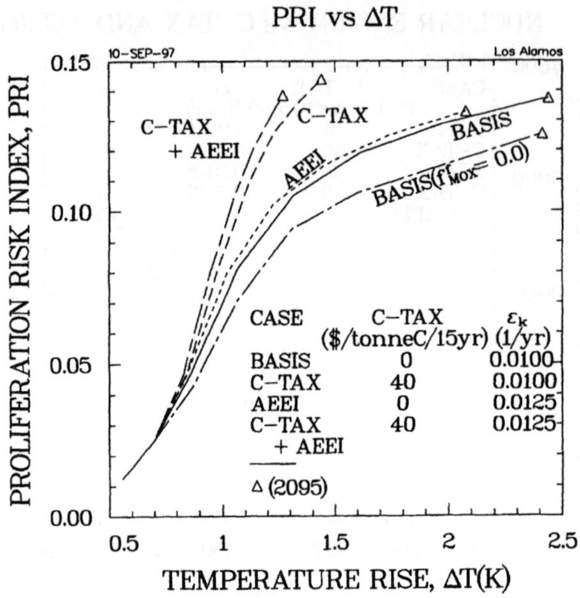

Figure 27 Direct comparison of proliferation-risk-index *versus* atmospheric temperature-rise "operating curves" for a combination carbon tax rate and AEEI, showing relative impacts of supply-side (carbon tax) and demand-side (AEEI) scenario attributes.

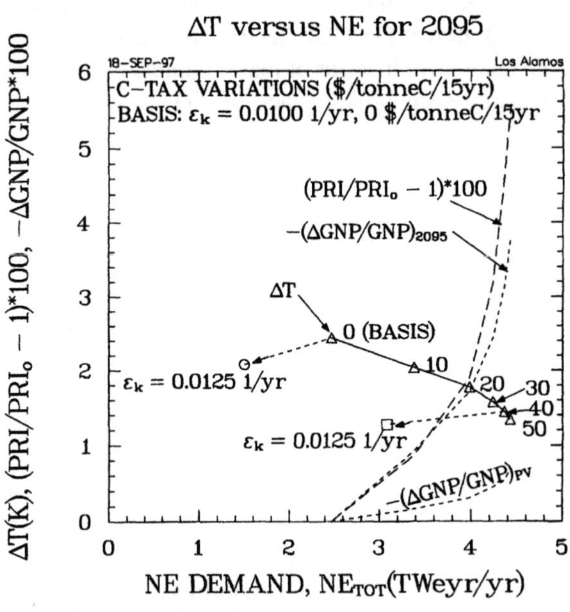

Figure 28 Dependence of global temperature rise on carbon-tax-induced (supply-side) nuclear-energy demand, showing impact of demand-side increases in AEEI for the Basis Scenario (no carbon taxes) and for a case where the carbon tax is imposed at a rate of 40 $/tonneC/15yr; also shown is the increase in PRI relative to the Basis Scenario as well as the relative decrease in GNP.

- *Non-Carbon Energy Sources:* The focus of this study has been on nuclear energy as a NC energy source, and even then only in so far as electricity generation is concerned; improved modeling of both solar and biomass[17] sources in the context of the present version of the ERB model is needed.

- *Nuclear Energy Model:* While attempts were made to introduce a "bottom-up" flavor into the nuclear model[6] used in the "top-down" ERB model, more remains to be done:[9]

  - *Nuclear Costing:* Attempts to fit a "bottom-up" feature in the costing of nuclear energy to the generically "top-down" ERB model need expansion to include more detail in both the fuel cycle and the capital cost inputs to the composite unit cost of energy used in ERB ultimately to determine nuclear energy market shares and related proliferation *versus* climate-change tradeoffs; central to improving fuel-cycle costing algorithms is the need to select regional and temporal plutonium recycle options based on economics rather than (region-dependent) exogenous drivers.
  - *Nuclear Materials Flows/Inventories:* While resolution into the range of forms (*e.g.*, reactor, spent-fuel, separated, *etc.*, plutonium) with which proliferation risks can be assessed is adequate, a rule-based algorithm for inter-regional transport and accumulations of plutonium based both on costs and sanctions needs to be developed to resolve and optimize local plutonium demand and supply; as noted below, consideration of both commercial Liquid Metal (Breeder) Reactors (LMRs) and LWR-supportive Fast-Spectrum Burners (FSBs) expands the scope of this issue.
  - *Breeder Requirements:* Integration of plutonium requirements of an evolving breeder economy *vis á vis* a coupling of regional and temporal breeding ratios to other parts of the nuclear fuel cycle is needed for any study that seriously evaluates and optimizes the potential and need for breeder reactors; the strong introduction of carbon-tax-induced nuclear energy, depending on which uranium-resource "reality" is adopted,[9,33] may advance the date for economic introduction of breeder reactors;
  - *Fast Spectrum Burners:* Comments made in connection with the last three items as related to improved understanding of the short- and long-term role of FSBs in closing the nuclear fuel cycle apply here also.
  - *Neutronics:* The neutronics model used to feed the nuclear materials flow and inventory model represents a highly approximate description of the time-averaged reactor core isotopics; the relationships between the many parameters listed on Fig. 3B and in Table IV of Ref. 9 need a firmer connection with "real" neutronics computations, particularly with regard to the averaged relationships between beginning- and end-of-life plutonium concentrations, overall fuel burnup, MOX core volume fractions, and fissions occurring in bred material.
  - *Fuel Cycle:* The impact of innovative/emerging fuel cycles (high-burnup, partial separations, non-aqueous processing, supportive transmutors, reactor integration, *etc.* on cost and proliferation risk needs detailed technological and economic assessment.

REFERENCES

1. E. D. Arthur and R. L. Wagner, Jr., "The Los Alamos National Laboratory Nuclear Vision Project," Proc. of the Uranium Institute 21st Annual Symposium, London, UK (September 4-6, 1996).
2. E. D. Arthur and R. L. Wagner, Jr., "The Los Alamos National Laboratory Nuclear Vision Project," Proc. Global Foundation Energy Conference, Miami Beach, FL (November 8-10, 1996), Plenum Press, New York, NY (1997).

3. P. Schwartz, The Art of the Long View: Planning for the Future in an Uncertain World, Doubleday Press, New York (1996).
4. H. McRae, The World in 2020: Power, Culture, and Prosperity, Harvard Business School Press, Boston, MA (1994).
5. J. Edmonds and J. M. Reilly, Global Energy: Assessing the Future, Oxford University Press, New York, NY (1985).
6. R. A. Krakowski, "Global Energy Modeling in Support of Understanding Long-Term Nuclear (Materials) Futures," Los Alamos National Laboratory document LA-UR-96-1931 (June 5, 1996).
7. R. A. Krakowski, "Global Nuclear Energy/Materials Modeling in Support of Los Alamos Nuclear Vision Project: Long-Term Tradeoffs Between Nuclear- and Fossil-Fuel Burning," Proc. Global Foundation Energy Conference: Technology for the Global Economic, Environmental, and Survival and Prosperity, Miami Beach, FL (November 8-10, 1996), Plenum Press, New York, NY (1997) [also, Los Alamos National Laboratory document LA-UR-96-4873(Rev.) (February 14, 1997)].
8. R. A. Krakowski, J. W. Davidson, C. G. Bathke, E. D. Arthur, and R. L. Wagner, Jr. "Global Economic/Energy/Environmental (E^3) Modeling of Long-Term Nuclear Futures," Proc. Global'97 Intern. Conf. on Future Nuclear Systems, Yokohama, Japan (October 5-10, 1997).
9. R. A. Krakowski, "Long-term Global Nuclear Energy and Fuel Cycle Strategies," Los Alamos National Laboratory document LA-UR-97-3826 (September 24, 1997).
10. R. A. Krakowski, "A Multi-Attribute Utility Approach to Generating Proliferation-Risk Metrics," Los Alamos National Laboratory document LA-UR-96-3620 (October 11, 1996).
11. P. Silvennoinen and J. Vira, "An Approach to Quantitative Assessment of Relative Proliferation Risks from Nuclear Fuel Cycle," J. Oper. Res., 32, 457 (1981).
12. P. Silvennoinen and J. Vira, "Quantifying Relative Proliferation Risks from Nuclear Fuel Cycles," Prog. Nuclear Energy, 17(3), 231 (1986).
13. I. A. Papazaglou, E. P. Gyftopoulos, M. M. Miller, N. C. Rasmussen, and H. A. Raiffa, "A Methodology for the Assessment of the Proliferation Resistance of Nuclear Power Systems," Massachusetts Institute of Technology report MIT-EL 78-021/022 (September 1978).
14. C. D, Heising, I. Saragossi, and P Sharafi, "A Comparative Assessment of the Economics and Proliferation Resistance of Advanced Nuclear Fuel Cycles," Energy, 5, 1131 (1980).
15. K. Hasselmann, S. Hasselmann, R. Giering, and V. Ocana, "Optimization of CO_2 Emissions Using Coupled Integral Climate Response and Simplified Cost Models: A Sensitivity Study," Climate Change: Integrating Science, Economics, and Policy, N. Nakicenovic, W. D. Nordhaus, R. Richels, and F. L. Toth, (Eds.) International Institute of Applied Systems Analysis report CP-96-1 (1995).
16. U. Cubasch, K. Hasselmann, H. Maier-Riemer, E. Mikolajewicz, B.D. Santer, and R. Sauser, "Time-Dependent Greenhouse Warming Computations with a Coupled Ocean-Atmosphere Model," Climate Dynamics 8, 55 (1992).
17. J. Edmonds, M. A. Wise, and D. W. Barns, "The Cost and Effectiveness of Energy Agreements to Alter Trajectories of Atmospheric Carbon Dioxide Emissions," Energy Policy, 23(4/5), 309 (1995).
18. B. Keepin and G. Kats, "Greenhouse Warming: Comparative Analysis of Nuclear and Efficiency Abatement Strategies," Energy Policy, 16(6), 538 (December 1988).
19. A. B. Lovins and H. L. Lovins, "Least-Cost Climatic Stabilization," Annual Review of Energy and Environment, 16, 433 (1991).
20. R. H. Williams, "Low-Cost Strategies for Coping with CO_2 Emission Limits," The Energy Journal, 11(3), 35 (1990).
21. A. S. Manne and R. G. Richels, Buying Greenhouse Insurance: The Economic Costs of Carbon Dioxide Emission Limits, MIT Press, Cambridge MA (1992).
22. P. Bagnoli, W. J. McKibbin, and P. J. Wilcoxen, "Global Economics Prospects: Medium Term Projections and Structural Change," Brookings Discussion Paper No. 121 in International Economics (January 1996).
23. J. C. Hourcade, "Modelling Long-Run Scenarios: Methodological Lessons from a Perspective on a Low-CO_2 Intensive Country," Energy Policy, 21(3), 309 (1993).
24. R. A. Krakowski, J. W. Davidson, C. G. Bathke, E. D. Arthur, and R. L. Wagner, Jr., "Nuclear Energy and Materials in the 21st Century, Intern. Symp. on Nuclear Fuel Cycle and Reactor Strategies: Adjusting to New Realities," IAEA, Vienna (June 3-6, 1997).
25. S. Messner and M. Strubegger, "Model-Based Decision Support in Energy Planning," International Institute of Applied Systems Analysis report WP-95-119 (November 1995).
26. K. Hanjalic (Ed.), Mathematical Modeling and Computer Simulation of Processes in Energy Systems, Hemisphere Publ. Corp., New York, NY (1990).
27. R. Bhatia, Econometric Methods for Energy Planning and Policy, World Scientific Publishing Co. New York, NY (1990).

28. E. R. Berndt and B. C. Field (Eds.) Modeling and Measuring Natural Resource Substitution, MIT Press, Cambridge MA (1981).
29. S. Messner and M. Strubegger, "User's Guide for MESSAGE III," International Institute of Applied Systems Analysis report WP-95-69 (1995).
30. G. A. Goldstein, "MARKAL-MACRO: A Methodology for Informed Energy, Economy, and Environmental Decision Making," Brookhaven National Laboratory report BNL-61832 (May 16, 1995).
31. J. F. Clarke, "The Cost and Benefit of Energy Technology in the Global Context," Proc. Conf. Technology Responses to Global Environmental Challenges: Energy Collaboration for the 21st Century, p. 521, Kyoto Japan (November 6-8, 1991).
32. W. Leontief, Input-Output Economics, Oxford University Press, New York, NY (1966).
33. "Uranium: 1995 Resources, Production, and Demand," Nuclear Energy Agency of the Organization for Economic Co-operation and Development (NEA/OECD) and the International Atomic Energy Agency (IAEA) joint report, OECD, Paris (1996).
34. W. Burch, E. Rodwell, I. Taylor, and M. Thompson, "A Review of the Economic Potential of Plutonium in Spent Fuel," Electric Power Research Institute report TR-106072 (February 1996).
35. M. Grubb, J. Edmonds, P. Ten Brink, and M. Morrison, "The Cost of Limiting Fossil-Fuel CO_2 Emissions," Ann. Rev. Energy Environment, 18, 397 (1993).
36. H. F. Wagner (Chairman), "Global Energy Outlook,"Symposium on Nuclear Fuel Cycle and Reactor Strategy: Adjusting to New Realities, Vienna (June 2-6, 1997).
37. N. Nakicenovic (Study Director), "Global Energy Perspective to 2050 and Beyond," World Energy Council (WEC) and International Institute for Applied Systems Analysis (IIASA) report (1995).
38. W. D. Nordhaus, "The Cost of Slowing Climate Change: a Survey," The Energy Journal, 12, 37 (1991).
39. C. Azar, "The Marginal Cost of CO_2 Emissions," Energy 19(12), 1255 (1994).
40. R. A. Krakowski, "Mitigation of Atmospheric Carbon Emissions Through Increased Energy Efficiency *versus* Increased Non-Carbon Energy Sources: A Trade Study Using a Simplified Market-Free Global Model," Los Alamos National Laboratory document LA-UP-97-3581 (September 5, 1997).

NOMENCLATURE

ACC	recyclable (to LWRs) accumulated plutonium; also, accelerator
AEEI	Autonomous Energy Efficiency Improvement
ATW	Accelerator Transmutation of Waste
BAU	business as usual
C-TAX	carbon tax
CAN	Canada
CHINA$^+$	China plus neighboring centrally planned countries
CIS	Commonwealth of Independent States (FSU)
COE(mill/kWeh)	cost of electricity
CON (EJ/yr)	consumption
CR	conventional uranium resources
D&D	decontamination and decommissioning
DEF	deforestation emission source
DEV	developing countries (ME + NAFR + SAFR + LA + IND + SEA)
DR(1/yr)	discount rate (for proliferation risk discounting,[11,12,10] or for estimating present worths of GWP or carbon taxes[38])
E^3	economics/energy/environmental
EEU	Eastern Europe
EI(GJ/$)	energy intensity, ratio of primary or final energy to GNP
ERB	Edmonds, Reilly, Barns model[5]
ES	energy services (residential/commercial, transportation, industrial)
FC	nuclear fuel cycle
FF	nuclear fuel fabrication
FP	fission product

FSB	fast spectrum burner (LMR/IFR, ATW)
FSU	Former Soviet Union
f_{MOX}	volume fraction of LWR core that is MOX
f_{MOX_f}	final (asymptotic) volume fraction of LWR core that is MOX
G(B$/yr)	gross world product, also GWP
GCC	global climate change
GDP($/yr)	gross domestic product
GHG	greenhouse gas
GNP($/yr)	gross national product
GWP($/yr)	gross world product
HV	high (nuclear energy growth) variant
HYDRO	hydroelectric
IAEA	International Atomic Energy Agency
IFR	Integral Fast Reactor
IIASA	International Institute for Applied Systems Analysis
IND	India
IPCC	Intergovernmental Panel on Climate Change
IRV	industrial revolution
j	ERB index for secondary energy (SE)
k	ERB index for energy services (ES)
LA	Latin America
LMR	liquid metal reactor
LV	low (nuclear energy growth) variant
LWR	light water reactor
KR	known uranium resources
LWR	light water reactor
l	ERB index for region
M_{CO_2} (Gtonne)	accumulated CO_2 emissions
$\dot{M}_{CO_2}$ (Gtonne/yr)	rate of CO_2 emissions
MAU	multi-attribute utility (analysis)
ME	Middle East
MM	(uranium) mining and milling
MIP	mixed integer programming
MIT	Massachusetts Institute of Technology
MOX	mixed (uranium, plutonium) oxide fuel
MV	medium (nuclear energy growth) variant
m	ERB index for time
N	number of MOX recycles; nonintervention scenario class
NAFR	Northern Africa
NE	nuclear energy
NM	nuclear materials
NT	no carbon taxes
NUCL	nuclear
nl	number of regions modeled in ERB (nl = 13)
O&M	operation and maintenance
OECD	Organization for Economic Co-operation and Development (USA + CAN + OECD-E + OECD-P)
OECD-E	OECD-Europe
OECD-P	OECD-Pacific
ORNL	Oak Ridge National Laboratory
OT	once-through (LWR)
P	parametric variation scenario class

P_E(MWe)	net electric generation capacity
P_{ET}(MWe)	total electric generation capacity
PE	primary energy [oil, gas, solids (coal + biomass), nuclear solar, hydroelectric]
POP	population
PRI	proliferation risk index
PRI_o	proliferation risk index without carbon taxes
PV	present value computed using discount rate DR
ppmv	volume parts per million (2.13 GtonneC/ppmv)
p_f	plant capacity factor
$R_{CO_{2},l}$(Gtonne/yr)	carbon emission rate from l^{th} region; for world total, (l = nl +1), same as $\dot{M}_{CO_2}$
REA	reactor plutonium
REC	fully recycle (N recycles to LWRs) plutonium
REF	(economically) reforming countries (EEU + FSU); also, reference time (1975)
RP	reprocessing
RPU	reactor plutonium
RS	repository
SAFR	Southern Africa
SE	secondary energy (liquids, gases, solids, electricity)
SE/PE	PE $\rightarrow$ SE conversion efficiency
SEA	South and East Asia
SEP	separated plutonium
SF	spent fuel
SFT	total spent fuel
SPU	separated plutonium (RP + FF)
TAX	carbon tax case designator
TH	thermal $\rightarrow$ electric conversion
TOT	total, world
TR	total uranium resources
TX	present worth of carbon taxes over period to 2095
t(yr)	time
t_{IRV}	time industrial revolution commences, 1800
t_{REF}	reference time or base year for ERB, 1975
USA	United States of America
UTC_j(\$/We)	unit total cost of jth nuclear energy system
<u>	grand utility function[10]
W(GtonneC)	atmospheric carbon-dioxide (carbon) accumulation
W_o(GtonneC)	integrated atmospheric carbon-dioxide (carbon) emissions, since t_{IRV} = 1800
W_{IRV}(GtonneC)	atmospheric carbon-dioxide (carbon) at time t_{IRV} = 1800 (594 Gtonne, or 289 ppmv, given 2.13 GtonneC/ppmv)[15]
We	electrical Watt
Wt	thermal Watt
WEC	World Energy Council
Z_{lm}	population in region l at time interval m
ΔT(K)	average global temperature rise, referenced to time t_{IRV} = 1800
ε_i(1/yr)	annual growth rate of entity i (i = POP, EI, PE, NE, *etc.*)
ε_k(1/yr)	annual growth rate of SE(j) $\rightarrow$ ES(k) transformation efficiencies
η_{TH}	thermal-electric conversion efficiency

PROSPECTS OF AND REQUIREMENTS FOR NUCLEAR POWER AS A CONTRIBUTOR TOWARD MANAGING GREENHOUSE GASES*

J. A. Hassberger, R. N. Schock, T. H. Isaacs

University of California
Lawrence Livermore National Laboratory
PO Box 808
Livermore, CA 94550

SUMMARY

The world's population, energy demand, and rate of carbon emissions are increasing. Even modest growth rates present significant challenges to existing and developing technologies for reducing carbon and greenhouse gas emissions while meeting growing energy demands. Nuclear power is currently the most developed alternative to fossil fuel combustion and is one of the options for meeting these challenges. However, there are significant technical, economic and institutional barriers inhibiting growth of nuclear capacity in the U.S. and slowing the rate of implementation worldwide. In the near-term, the major barriers to nuclear power, especially in the U.S., appear to be economic and institutional. Risks such as safety, waste management and proliferation are within reasonably acceptable limits at the current installed capacity. Future growth of nuclear power, however, may well hinge on continuous evolutionary and perhaps revolutionary improvements in safety and fuel cycle management so that the overall risk of nuclear power, aggregated over the entire installed capacity, remains at or below today's risks. This is a challenge that both U.S. industry and the government must step up to with a well-focused R&D program.

INTRODUCTION

Nuclear energy is one of the current contributors to, and future options for, reducing carbon and greenhouse gas emissions. As a demonstrated, safe and environmentally responsible source of electricity, nuclear power already avoids over 140 million metric tonnes of carbon (MtC) emissions annually in the United States, representing a 10% re-

* This work was performed under the auspices of the U.S. Department of Energy by the Lawrence Livermore National Laboratory under Contract no. W07405-Eng-48.

duction in U.S. annual emissions, and some 500 MtC annually worldwide. Nuclear plants can provide further reductions by replacing fossil fuels in electric production, and has potential for more reductions by providing process heat for industrial applications, district heating and alternative transportation fuels manufacturing.

The nation's energy demand is increasing, the world energy consumption rate is increasing even faster, and throughout the world, electricity's contribution to total energy consumption is rising. Increased demand is only part of the driver for new generating plant construction, as new plants are also needed to replace aging and outmoded plants, both nuclear and fossil-fueled. One-third of the total new domestic generating plant construction requirements projected through 2015 is for plant replacement, not for demand growth.

The world's energy future is uncertain, with significant uncertainties in growth rate, technology availability, and economics. The distributions of energy demand growth and resources are disparate: growth is more rapid among the developing countries and especially in East Asia, and demand growth is not well matched to energy resource distribution. This leads to social disparities and the potential for mass migration and conflict, making energy security and sustainability national and international security issues. Finally, there is uncertainty about the world's resolve to effectively limit carbon and greenhouse emissions, especially in light of anticipated costs and impacts on economic growth. Without concurrent policies for emissions-free generation, or emissions based taxes on energy generation and use, carbon emissions in the U.S. and most other nations will continue to rise.

Flexibility and diversity are the keys to ensuring a responsible energy future in light of all these uncertainties. Ensuring a safe, economic and reliable energy supply to meet anticipated future demands must be high priority national and international goals. Technology flexibility and fuel diversity is needed to guard against unforeseen shortcomings and supply disruptions. A fully integrated mix of energy supply technologies can effectively maximize the benefits of each technology while minimizing the risks and costs of our energy supply systems as a whole.

Today's Energy and Carbon Emissions Spectra

Energy generation and use accounts for the vast bulk of carbon emissions. In the U.S. alone, the energy sector contributes some 90% of the nearly 1500 MtC emitted annually. The transportation, industry and building (air conditioning, heat and light) sectors each contribute roughly 30% of the carbon emissions.

Much of these emissions are avoidable. Today's carbon emissions would be 15% higher if it were not for emissions avoided through nuclear power (140 MtC/a, 10%) and renewables (primarily hydro, 70 MtC/a, 5%). Ultimately, a majority of current and future emissions could be avoided through a combination of renewable and nuclear energy sources combined with substantial shifts in infrastructure and energy transport systems, albeit on varying time frames.

Fossil electrical sources account for some 30% of current carbon emissions. In principal, most of this generation, especially that used for baseload, can be replaced with a combination of nuclear and renewable sources. This is a proven capability of nuclear power today.

Transportation energy use accounts for another 30% of current carbon emissions, with perhaps half of this from the light vehicle fleet, including passenger cars and light trucks. There are reasonable prospects that a substantial portion of the light fleet emissions can be eliminated by conversion to electric vehicles and hybrid-electric vehicles operating on hydrogen. Notwithstanding the need for continued development of the vehicle technologies and hydrogen infrastructures, this can be considered a current capability of nuclear as a primary electrical generator, and a potential future capability using hydrogen produced with nuclear process heat.

Some of the remaining emissions, roughly 10% of the total annual (162 MtC/a) is used for building heat and 20% (382 MtC/a) is used in industry, substantially for process heat. Some fraction of these emissions could be eliminated using electric heat from nuclear or renewable sources. Both nuclear process heat and nuclear district heating has had limited demonstration internationally, and could be considered a potential capability in the U.S.

The World's Energy Future

Trends in the world's population and energy use over the past century show dramatic increases in both. The beginning of the century saw a world population of only some 1.5 billion. It is now fast approaching 6 billion. As late as 1950, the world's annual consumption of primary energy barely exceeded 100 exojoules (EJ, 1 EJ=10^{18} Joules). Today, it is on the order of 400 EJ per year. Energy use is projected to increase at an even more rapid rate. As Figure 1 shows, even though various projections of population and energy growth vary, even the slowest growth scenarios offer exceptional challenges for the future. By 2020, population is expected to be at least 7.5 billion, and consume between 600 and 750 EJ per year.

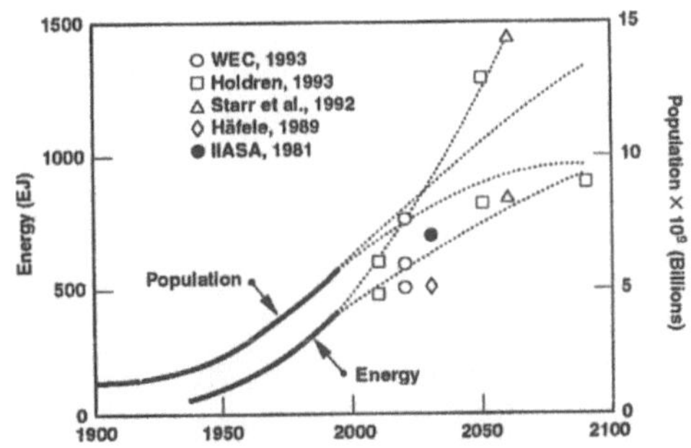

Figure 1. Actual (to 1996) and projected global population growth and energy usage[1].

Electrical consumption is expanding much more rapidly than total energy use. The EIA projects worldwide consumption of electricity to increase over 4-fold, while total energy consumption increases "only" about 2.5 times. This trend is even more dramatic among the developing countries, where total energy consumption for the same period is expected to increase over 6-fold, but electrical consumption is expected to increase over 14-fold.

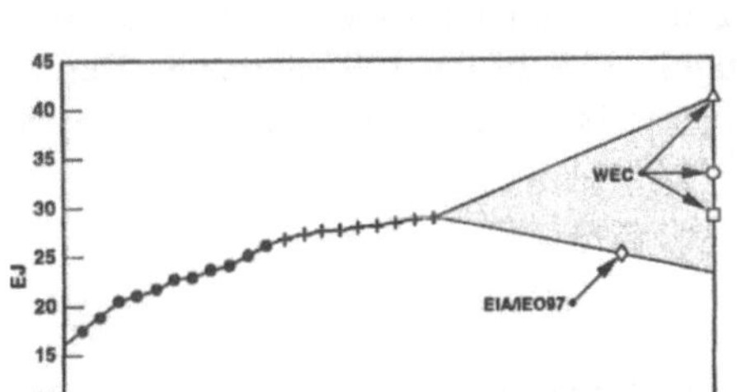

Figure 2. Actual (to 1996) and Projected global nuclear generating capacity (EJ/year)[2].

Projections for the future of nuclear power vary greatly, but most show significant growth for nuclear power everywhere except in the U.S. Globally, most projections show slight to moderate growth for nuclear power, but include projected declines in U.S. domestic nuclear power generation. Figure 2 summarizes the historical capacity, capacity under construction and projections for global nuclear energy through 2020. The only case suggesting a worldwide decline in nuclear energy is the IEO97 reference case, projecting nuclear

[1] Energy scenarios encompass high and low ranges and are from: World Energy Council (1993) for 2020 (three cases - high growth, reference, ecologically driven); Holdren (1992) for 2010, 2050 and 2090; Starr, et al. (1992) for 2060; Hafele (1989) for 2030; and IIASA (1981) for 22030. The World Bank estimates (1994) a population of about 11 billion in 2100. Low energy estimates assume, among others, continuous energy efficiency improvements on the order of 50 percent by 2100.

[2] U.S. Energy Information Administration, 1997

capacity to decline from 344 gigawatts in 1995 to 332 gigawatts in 2015, primarily because nearly one-third of the present U.S. nuclear capacity (100 Gwe) is scheduled for retirement by 2015[3]. While not enough to offset planned retirements, significant increases in nuclear capacity are projected for developing countries, including an additional 31.5 gigawatts by 2015 in China, South Korea, India, and Taiwan. This area has the highest rate of growth in electricity demand in the world and is therefore in need of extensive baseload capacity additions. Current and projected regional distributions of nuclear generating capacity under this scenario are summarized on Figure 3.

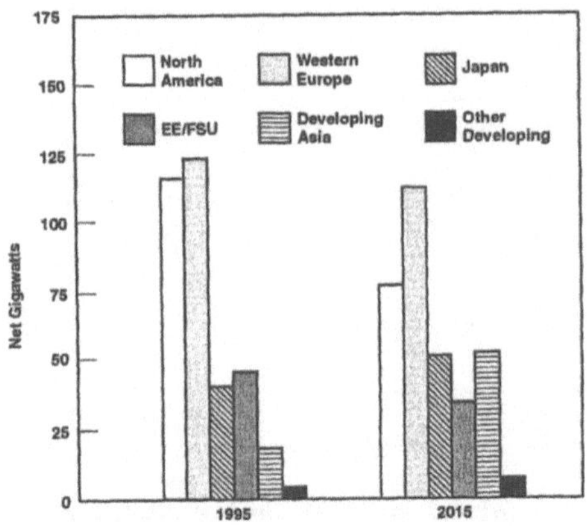

Figure 3. World nuclear energy capacity by region in 1995 and 2015[4].

Much attention has been given to anticipated retirements of aging and uneconomic nuclear plants, and the adverse impact replacing these plants with fossil-fueled plants would have on carbon and greenhouse gas emissions. In addition to the 38 gigawatts of retired domestic nuclear capacity in the AEO97 case[5], the EIA anticipates 71 gigawatts of fossil generation will also retire by 2015. Combined with new growth demand, a total of nearly 320 gigawatts of new generating capacity will be needed in the U.S. by 2015, representing construction of nearly 1100 gas-fired generating plants, or over new domestic 50 plants per year. Considering sensitivities to carbon emissions, implications of deregulation, and costs and delays likely resulting from local opposition to new plant construction, life extension of existing nuclear plants could be most advantageous to assuring a stable and reliable electrical supply system.

Current EIA projections do not include projections for impacts of shifting transportation to electrical power, hydrogen or other manufactured fuels. Electrical demand would increase dramatically if significant fractions of transportation energy use were tied to it. For example, if only 20% of the estimate 180 million U.S. automobiles and light trucks were converted to electrolytically produced hydrogen fuel, annual domestic electrical demand would increase by over 400 billion kWh, roughly 12% of the 1995 demand. Berry, et al. (1966) have shown that such vehicles can be operated competitively with today's vehicles using electricity obtained from the grid.

---

[3] Current discussions about life extensions could reduce or delay the projected decline in U.S. nuclear capacity and reduce the need for replacement capacity.

[4] International Energy Outlook, U.S. Energy Information Administration, 1997. The EIA high-growth case projects nuclear power generation to increase from the 1995 capacity of 344 Gwe to nearly 450 Gwe by 2015. This high-growth case projects only a small decline in domestic U.S. nuclear capacity (from 100 to 95 Gwe) over this period, but a substantial growth in nuclear capacity in essentially all other parts of the word: nuclear power in the rest of the industrialized countries grows from 178 to 212 Gwe; Eastern Europe and FSU from 46 to 66 Gwe; the developing countries from 21 to 77 Gwe.

[5] Annual Energy Outlook, U.S. Energy Information Administration, 1997

An uncertain future

The world's energy future is uncertain, and there are disparities in the distributions of energy demand growth and resources. Growth is more rapid among the developing countries and East Asia, and demand growth distribution is not well matched to energy resource distribution. Shifts among various energy sectors, such as transportation, are likely. However, such shifts, as well as our future generation mix, will depend on availability of new or improved technologies. Future energy economics are clouded by deregulation and the potential impact of difficult decisions on environmental issues and policies.

The world's fossil fuel resource base, outside of coal, is limited. Coupled with disparities in distribution, this is a cause of global concern for energy sustainability and energy security. Economics, jobs, standards of living and birth rates are inexorably tied to the availability of affordable energy. Energy disparities lead to economic disparities and potential for mass migration and regional conflict, making energy security and sustainability national and international security issues.

There is also great uncertainty surrounding the technologies and technology mix used to meet this demand. Alternative and renewable technologies are in the same romantic period of development that nuclear fission and fusion enjoyed in the 60s, and space flight in the 70's. As in other historical technology development cases, projections, promises, and hopes fell short as physical laws, limitations of materials and infrastructures, energy demand, and increasing public sensitivities failed to meet expectations. The reality of any energy technology development is that it takes far longer to reach its perceived goals than anticipated. Commercial development is best planned based on the results of technology demonstrations rather than hopes.

The impact of utility deregulation on both nuclear and renewable energy technologies is expected to slow the growth of these sources without some form of strong incentives. Wind power is projected by the EIA as the major non-hydro renewable resource likely through 2015. However, even in regions with favorable sites, the EIA notes that in an uncertain (deregulated) power market, without incentives (such as a carbon tax or carbon emissions cap) wind power will likely be used primarily as a fuel-saver and its overall contribution limited. The cost of solar photovoltaic is expected to remain too high for use in centralized power stations, and solar PV will likely remain an affordable option primarily for remote power generating applications.

Finally, there is uncertainty about the world's resolve to effectively limit carbon and greenhouse emissions, especially in light of anticipated costs and impact on economic growth. In the U.S. deregulation is moving the nation towards lowest cost options, which are expected to result in ever greater reliance on fossil fuel generation. Without specific incentives for emissions-free generation, or emissions based taxes on energy generation and use, carbon emissions in the U.S. and most other nations will continue to rise.

All of these uncertainties, the global energy future, the state of various energy technologies, and the lag time between policy decisions and market impact make it imperative that we maintain a robust set of options for meeting our current and future energy demands in environmentally responsible ways. Nuclear power, as a proven resources, is one of these options.

BENEFITS AND RISKS OF NUCLEAR POWER

Nuclear power is an important, and sometimes contentious, national issue. It has both risks and benefits, just as any other technology. Unlike some national issues, however, there are important risks that accrue in today's world if the nuclear power option is not pursued.

Some of the major benefits of nuclear power are:

- It is a significant, stable and independent domestic and international energy resource, offering needed diversity and flexibility against future uncertainties,
- It is safe and environmentally sound, emitting few pollutants and no carbon or other greenhouse gases,
- Nuclear power is an important source of U.S. high-technology exports, and foreign demand for such exports is increasing,
- Nuclear power maintains U.S. positions of technical leadership and influence in important international policy areas such as nonproliferation, safety and waste management.

Even though most of the risks associated with nuclear power have, or are ammenable to, technical solutions, there are significant institutional, political and social acceptance issues associated with nuclear power:

- There are proliferation implications and perceived risks with nuclear power internationally,
- Spent fuel and waste management are significant institutional issues,
- Consequences of a nuclear accident remain a significant issue,
- Nuclear power cost is currently high relative to most alternatives.

Just as there are risks associated with nuclear power, there are risks associated with *not* exploiting nuclear power:

- Loss of international influence and technical leadership,
- Erosion of critical infrastructures necessary to effectively deal with significant nuclear issues, including D&D, waste management, and safety,
- Impaired energy options and flexibility.

If the U.S. is going to meet its international environmental commitments, support a growing economy, and ensure its own as well as others' energy security, the nation must find ways to enhance and support nuclear energy as a part of the U.S. energy portfolio. The significant energy security and environmental benefits of nuclear power will only be realized if the undesirable risks associated with nuclear power are reduced. These risks can be reduced by well-focused research and development.

In light of major uncertainties in the world's energy outlook and markets, clear needs for facilities and expertise to deal with existing and future security, safety and environmental issues, and continuing international tensions in much of the world, the U.S. simply cannot afford to allow its nuclear influence, infrastructures or options to erode further.

THE ISSUES

There are four issues dominating public debate over nuclear power: nuclear safety, waste and spent fuel management, nuclear proliferation and economics. The risks associated with each of these issues vary widely depending on one's perspective. Proponents of nuclear power point out, with good justification, that nuclear power is already much safer that the alternatives, waste and spent fuel management are technically solved issues, and observe that no nation today has used safeguarded commercial facilities to develop nuclear weapons. Opponents, relying substantially on the linear no-threshold model for radiation exposure, see the consequences of even low-probability events and the long confinement times needed for waste and spent fuel disposal as problematic, and fear the potential for technology diffusion and materials diversion from even safeguarded facilities. All agree that the cost of nuclear power is too high, but disagree on the reasons for the costs and the methods for reducing them.

With these disparate perceptions, it is difficult to quantify these risks, and even more difficult to arrive at a consensus definition of acceptable risk. One observation is clear, however: a combination of real risk reduction (through R&D and technology implementation) along with an improvement in risk perception (through institutional, policy and regulatory reform, and shifts in public perception) in necessary to improve the acceptability of nuclear power today.

There are R&D roles that both government and industry can and should play. Government's role in energy R&D is to help prepare the country for probable futures and guard against improbable ones. This means ensuring that the technical bases for a wide variety of energy options are available to meet the upcoming challenges of our future energy needs. Industry's role is to take these technical bases and convert the most appropriate and economic options into industrialized and commercialized systems.

The first three of the four issues, nuclear safety, waste and spent fuel management, and nuclear proliferation, are unique to nuclear power. They can be addressed essentially independent of the alternatives. Economics, on the other hand, is not independent of the alternatives. All of the four issues are both technical and institutional issues. Nuclear power offers lower technical risks than the available alternatives, technical solutions for waste management issues exist, and the LWR fuel cycle without reprocessing is quite proliferation resistant. Recent construction of the Advanced Boiling Water Reactor in Japan, accomplished in only four years, demonstrates that the technology for helping reduce costs in also in place. Thus, these issues are substantially institutional in nature. Even so, continued technological improvements can have major impacts on their ultimate risks.

In addition to these four issues that must be addressed if nuclear power is to have a significant place in the U.S. energy future, there are very difficult challenges that must be met even if nuclear power is allowed to decline. The current trend in the U.S. is abandonment of nuclear power in the face of global growth. This current trend presents the U.S. with three important challenges:

- *Influence:* How does the U.S. best influence the rest of the world in critical nuclear issues such as nonproliferation, waste management, and nuclear safety?
- *Infrastructure:* How do we maintain sufficient expertise and infrastructure to deal with the enduring nuclear issues such as waste and safety, both domestically and internationally?
- *Future Options:* Can the U.S. truly afford to preclude the use of nuclear energy in a future with uncertain environmental and energy security issues?

Meeting these challenges will require both domestically and internationally focused efforts. Domestically, significant continuing research is needed on safety, waste management, and spent fuel minimization, and real efforts are needed focused on infrastructure maintenance, including preservation of critical educational and research resources. As a nation, we cannot afford to let our technical leadership and influence in important international nuclear issues wane. Nuclear proliferation, and international nuclear safety are issues in which the U.S. is losing, or threatening to loose, its influence because of its current policies and decline of its nuclear industry. Both the U.S. and the rest of the world have an enduring need to maintain vital infrastructures to deal with nuclear waste, ensure continued safe and reliably operation of all nuclear plants, and to deal with the legacy of nuclear weapons production. A well-developed nuclear R&D program is essentially for meeting these vital, enduring national interests.

Nuclear Safety

Safety is simultaneously a source of nuclear power's benefits and risks. Nuclear power is operationally as safe as, or safer than, most alternatives (coal plant effluents cause cancer, dams have failed). However, the consequences of a serious accident are potentially large. While most Americans today appear to accept nuclear power as sufficiently safe, they are concerned about accident consequences and are unwilling to pay a premium to further reduce risks. Thus, safety is intimately tied to economics and remains an important issue. Certainly the events at Three Mile Island and Chernobyl have made nuclear safety an enduring international nuclear issue, with events now having global impacts. Continued R&D on reactor safety has proven beneficial, as the Advanced Light Water Reactor program has demonstrated. Further improvements in the safety of both existing and evolutionary reactor systems, as well as development of reactor systems featuring even greater reliance on inherent safety (rather than technical safety) will be key to the continuation of nuclear power as a viable energy alternative.

Waste Management

Waste Management, particularly that of spent fuel is one of the more contentious concerns the U.S. public has regarding continued use of nuclear energy. The Federal government is having difficulty meeting its statutory responsibility to manage the spent fuel from the nation's commercial nuclear power plants, leading to frustration on the part of utilities, electricity customers, and state leaders. Although reasonable, sound technical solutions to this issue are essentially on-hand today, very vocal demands for non-attainable repository standards make waste management a thorny political and policy issue. Even so, there are technology avenues that can result in reduced waste and spent fuel accumulation, and continued R&D in this area can have important impacts.

Nuclear Nonproliferation

Nuclear proliferation is one of the risks associated with nuclear power. Reducing the risk of proliferation is and will remain an important worldwide goal, and it is the motivating factor behind much of the nation's international nuclear dialogue and diplomacy. Two issues underlie proliferation risk: the illegal or clandestine use of declared facilities and/or materials, and theft of weapons-useable materials. Safeguards are an effective deterrent against diversion and misuse, and the major theft-related proliferation risks arise from commercial nuclear activities including plutonium trade and transport associated with plutonium recycle and MOX utilization. This risk is exacerbated by the growing stocks of separated plutonium due to the imbalance between separation and utilization. Large commercial reprocessing activities are in place in Europe. Russia, Japan and China have commercial reprocessing facilities under construction or planned. Other countries are likely to pursue reprocessing, both as a mechanism for extending energy resources and for waste management. Research into new fuel cycles for reduced spent fuel accumulation and reduced plutonium buildup could have substantial favorable impact on proliferation resistance. So could research into new processes for proliferation-resistant waste isolation

Economics

Economics of nuclear power in a changing domestic energy market is an increasingly important issue. Deregulation, with its attendant focus on lowest short-term generating costs, is disadvantageous to capital-intensive installations such as nuclear plants (and to large coal-fired plants as well). Long construction times increases both cost and financial risk of nuclear projects, and although recent regulatory reforms have helped to address this issue, more streamlining is needed.

Many of nuclear power's environmental costs have been internalized, through assessments to pay for spent fuel disposition. Currently, fossil environmental costs have not been internalized, a situation viewed by many as resulting in unfair economic competition. The Interlaboratory Working Group (1977) notes that the cost of stabilizing U.S. carbon emissions at 1990 levels by the year 2010 has been estimated to require a carbon tax equivalent to somewhere between $10 and $130 per ton of carbon, and cites another study suggesting that carbon emissions avoided might be valued at $50/ton. Using this later value, and a value of 0.26 MtC/TWh for fossil fuel electrical generation suggests these cost might be equivalent to some 1.3 cents/kWh. Imposition of such an assessment (similar to that imposed on nuclear for spent fuel management) on fossil fuel generation would make both renewables and nuclear comparatively more attractive.

PROSPECTS AND REQUIREMENTS FOR A NUCLEAR FUTURE

Even though nuclear power has obvious and substantial advantages as an effective tool for managing carbon and greenhouse gas emissions while meeting energy demands, its economic costs are too high for substantial near-term growth in the U.S. For the longer term, the unique nature of nuclear power risks, coupled with unrealistic public perception of these risks is a significant impediment. Long-lived radioactive waste, reactor safety, and nuclear proliferation make a fundamental difference in the way associated risks are perceived and evaluated. Ultimately, the future of nuclear power depends on our ability to continue to reduce the risks of nuclear power, both relative to the alternatives, and to absolute levels palatable to the general public.

One of the perceived differences between nuclear power and its alternatives may be that nuclear events are seen as having direct, global impacts, whereas events at conventional power plants are viewed as local. In terms of risk, the public therefore tends to treat nuclear risks as aggregated — an event anywhere in the world matters. This has significant implications for the future of nuclear power.

Consider the implications for reactor safety. Current reactors are designed to a goal of less than one chance of a serious accident in a million reactor-years. The current U.S. reactor fleet of some 100 reactors therefore represents a risk of a serious accident of roughly one in ten thousand years. The general public appears to accept this level of risk, as it is not clamoring for abandoning existing plants, but neither is it tolerant of new nuclear construction. Thus, we might take this level of risk to represent a limit for public acceptability of nuclear risks. Therefore, for an increase in the number of reactors in the U.S. to be publicly acceptable, safety standards, designs and practices themselves may require improvements to maintain this limit on aggregated risks.

Indeed, such an approach is neither unreasonably nor unprecedented. Even though our automotive travel increases, the U.S. public continues to demand, and receives, improvements in safety as measured in per-mile risks, thus maintaining control of overall aggregated automotive risks.

Engineered approaches to nuclear safety (and other risk reduction) may be reaching a point of diminishing returns, both technically and in the public's perception. Both the Three-Mile Island and Chernobyl events happened substantially because of operator errors, and technical complexity is perceived to have contributed to these errors. The public sees the systems as too complex even for the operators to understand, and is wary of additional reliance on technological fixes. Efforts to develop "passively safe" systems is a step in the right direction, and one that should be extended to the other elements of the fuel cycle.

Similar factors may play in the perception of other nuclear risks, such as the risks of waste and spent fuel management. If we assume the perceived risks of these issues are similarly aggregated, then future growth of nuclear power depends on proportionate reductions in the risks as system capacities are increased. The current U.S. philosophy of direct disposal contributes to the aggregation of risk. Waste segregation, coupled with actinide burning have the potential for reducing or eliminating the bulk of the long-lived radionuclides from the waste stream and reducing repository performance requirements Variations on these systems offer potential for further reductions in both high-level waste activity and volume. Such simultaneous reductions in waste activity, volume and lifetime may be necessary for real future growth in nuclear power.

This view of risk highlights several significant challenges for the future of nuclear power. Continuing evolutionary improvements in safety, waste management and other issues appear sufficient to maintain the existing nuclear fleet and, given favorable economics, to perhaps support incremental increases in installed nuclear capacity. However, reemergence of nuclear power as a viable, growing energy option for the future may require new approaches to nuclear fission, including a thorough reassessment of the entire nuclear fuel cycle.

Maintenance of a viable nuclear option for the future will depend on how the U.S. nuclear community, its infrastructures, expertise and industry, weather the current interregnum in reactor design and construction. There is no current U.S. market for new nuclear generating capacity, and "optimistic" scenarios suggesting that the current nuclear share of the market be maintained (at roughly 20% of U.S. generating capacity) translates to only one or two 1300 MWe unit each year (with a current installed capacity of 100 Gwe and a projected electrical growth rate of some 1.5 %/year as suggested by the EIA reference case, and including replacements for retiring units.) Even this domestic market is likely insufficient to maintain effective U.S. nuclear infrastructures and industry. Rather, maintaining a viable domestic nuclear option for the future is intimately tied to the global nuclear market and an effective U.S. presence in that market.

Maintaining the U.S. presence in the international nuclear markets is itself a challenge, with serious competition from the French and Japanese, and a growing presence from others including Canada and Korea. This challenge is intensified by the coupling between U.S. policy and diplomatic concerns with international nuclear trade, as exemplified in the ongoing U.S.-Chinese nuclear trade negotiations.

Differences between U.S. nuclear policies and those of potential customers is also often problematic. An example is the U.S. position on reprocessing, which is seen as a deterrent to reliance on U.S. technologies and systems when potential customers view reprocessing as an important option for waste and resource management. To mitigate U.S. concerns about nuclear proliferation, progress is needed in both technological and institutional approaches toward reducing the risks associated with reprocessing. Technologically, development of fuels for extended burnup may help reduce waste generation and resource utilization, reducing the imperative for reprocessing. Development of better reprocessing technologies, with enhanced proliferation resistance could make reprocessing more acceptable. Institutionally, better international control of sensitive fuel cycle processes and facilities, through enhanced oversight, regionalization, or bi- or multi-lateral control could be particularly effective.

The market for nuclear power in the less developed countries is potentially large. However the ability of nuclear power to penetrate these markets will be limited by several of the issues already discussed, including cost and proliferation risk, as well as some factors unique to the less developed countries, such as lack of supporting infrastructures and expertise, and limited capacity to absorb current large capacity plants. Indeed, in the near term (through, say 2010) the global market for large power systems may be insufficient to support all existing reactor vendors. The ability of nuclear power to play a significant role in these countries will depend, in part, on the availability of nuclear systems capable of overcoming these limitations.

Less expensive, smaller systems, relying on passive safety designs, and featuring enhanced proliferation resistance (perhaps by avoiding on-site refueling through use of lifetime cores) could make nuclear power an attractive option for simultaneously meeting energy demands and managing greenhouse gas emissions in many less developed countries, especially those deficient in indigenous energy resources.

To maintain a presence in the international nuclear market and influence in important international nuclear issues, particularly among the developing countries, the U.S. must be responsive to market demands and the needs of the other countries. International dialogue and cooperation is essential for understanding the issues, capabilities and limitations of nuclear power from the perspectives of the other countries. Technical and institutional solutions to significant nuclear concerns that are effective in the U.S. or other industrial countries may be inappropriate for application in the less developed countries.

SUMMARY AND CONCLUSIONS

In this paper we have reviewed the four major impediments hindering a resurgence of nuclear power in the U.S., and to a lesser extent, in much of the world: economics, safety, waste and spent fuel management, and nuclear proliferation. We have identified three significant challenges for the nation that will become increasingly difficult if nuclear power is allowed to continue its current decline in the domestic energy market: loss of international incluence and technical leadership, erosion of critical nuclear infrastructures, and impairment of flexible energy options to meet growing energy and environmental demands.

The future of nuclear power in the U.S., notwithstanding the significance of the other major issues, is dominated by the currently excessive costs. As long as there are economically viable alternatives, the public appears unwilling to accept additional nuclear power risks. The economic future of nuclear power is further threatened the potential inability of the global market to sustain the current industrial base.

The future of nuclear power in the U.S. is, therefore, conditioned on two factors. Domestically, continued progress on cost and risk reduction in each of the major issues is necessary. Globally, the health of the U.S. nuclear industry is intimately tied to the global market, and depends on the U.S. ability to effectively compete in and influence that market.

Meeting the challenges facing nuclear power requires a well-focused R&D program to continue the evolutionary and perhaps revolutionary improvements in the safety and waste management issues. Developing and maintaining the global market needs continued progress in reducing proliferation concerns, and improved safeguards and security. Novel approaches to new systems focussed on non-traditional siting considerations in less developed nations could produce new markets and help avoid potential loss of domestic scientific and technical infrastructures and industry base.

REFERENCES

Berry, G.D., et al., 1996, Hydrogen as a future transportation fuel, *Energy* 21:4, 289-303

Häfele, W., 1989, Energy Systems under stress, World Energy Conference, Montreal, September

Holdren, J.P., 1992, Talk At Lawrence Livermore National Laboratory, February 19, 1993

International Institute for Applied Systems Analysis (IIASA), 1981, *Energy in a Finite World: Paths to a Sustainable Future*, Cambridge, MA; Ballinger Publishing Co.

Interlaboratory Working Group, 1997, *Scenarios of U.S. carbon reductions:potential impacts of energy technologies by 2010 and beyond*, LBNL-40533, Lawrence Berkeley National Laboratory

Starr, et al., 1992, Energy sources: a realistic outlook, *Science* 256, 981-7

U.S. DOE National Laboratory Directors, 1997, *Technology opportunities to reduce U.S. greenhouse gas emissions*, Draft, November 13, 1997

U.S. Energy Information Agency, 1997, *International Energy Outlook, 1997*, DOE/EIA-0484(97), United States Department of Energy

U.S. Energy Information Agency, 1996, *Annual Energy Outlook, 1997, with projections to 2015*, DOE/EIA-0383(97), United States Department of Energy

World Bank, 1994, *Population and Development: Implications for the World Bank*, Washington D.C.; The World Bank

World Energy Council, 1993, *Energy for Tomorrow's World*, New York, St. Martin's Press

PRINCIPLES OF QUALITY FOR NUCLEAR TECHNOLOGY: ESSENTIAL FOR SUSTAINABLE ENERGY AT REASONABLE COSTS IN A CLEAN ENVIRONMENT

Clinton Bastin

Chemical Engineer, United States Department of Energy (Retired)
Immediate Past President, National Treasury Employees Union Ch. 228
Signatory and a Director of the Eagle Alliance[1]
14355 Long Channel Drive, Germantown, Maryland 20874

INTRODUCTION

At a time of almost certain need because of local and global challenges to the environment from overuse of fossil and other organic fuels, nuclear technology may not be available in the United States because it lacks credibility with many Americans. It lost credibility because leaders of the nuclear community failed to ensure that important principles of quality were used to manage, regulate, and develop policies for its use.

Credibility can be restored, and nuclear technology can meet the challenges without major economic disruption. But to do so nuclear program, regulatory, and policy leaders must adopt the principles of quality that led to success, and avoid the practices that led to failures.

Table 1, a list of "Principles of Quality for Managing, Regulating, and Developing Policies for Use of Nuclear Technology," was developed from work with corporations in the U.S. and other nations that achieved success in nuclear programs through use of quality principles, and assessment of failures where quality principles were not used. It also incorporates ideas of National Treasury Employees Union (NTEU) President Robert M. Tobias for employee empowerment for quality and safety through partnerships between labor and management. Employee empowerment by The DuPont Company for input to managers on safety concerns was a major factor in achievement of a world-best safety record in use of nuclear technology at the Savannah River Plant for the U.S. Atomic Energy Commission (AEC).

---

1 The Eagle Alliance is a partnership agreement of individuals and representatives of corporations, universities, unions and other organizations who have worked to develop peaceful uses of nuclear technology and believe that nuclear technology is a proper, safe, and essential element of advanced civilizations. It is also a not-for profit corporation whose mission is to provide accurate information to America's citizens and leaders about the great benefits of nuclear technology, and ideas on how this technology can be better managed to provide these benefits, minimize potential for danger, and avoid wasteful expenditures of money.

Table 1. Principles of Quality for Managing, Regulating and Developing Policies for Use of Nuclear Technology

This list was developed from the author's work with organizations that achieved success with nuclear technology through use of quality principles, and assessment of problems and failures where good quality principles were not used. It also incorporates ideas of National Treasury Employees Union President Robert M. Tobias for empowerment of employees for quality through partnerships between labor and management.

1. Listens to, seeks to understand, and communicates with its employees, customers, suppliers, and the public; ensures full and flexible communication within the organization; ensures that full and accurate information is provided to the public, public officials, and the news media about its activities; and makes strong effort to immediately correct misinformation.
2. Ensures that there is good leadership. "Characteristics of Extraordinary Leadership" are discussed in Table 2. A critical need for energy and environmental program leaders is a full understanding of the laws of thermodynamics and principles of nuclear technology.
3. Ensures that full input is obtained from persons that understand details of technology being considered for use, and that design and operating details are included in evaluations.
4. Ensures that there is safety (including safeguards, fire prevention, and environmental and radiation protection) oversight by competent persons with global awareness of potential safety challenges and protection systems, but recognizes that safety is the primary responsibility of and is achieved by those performing the work.
5. Recognizes the value of strength through diversity and ensures that the diverse strengths from differing cultures are effectively utilized. The greatness of America has been achieved through the strengths of its different cultures.
6. "Sticks to the Knitting." This chapter of the book *In Search of Excellence*, by Thomas J. Peters and Robert H. Waterman, Jr., points out the difficulties of diversification of a corporation into areas where there is no corporate expertise.
7. Ensures that workers and managers form partnerships and interact as partners, that employees accept responsibility for and are proactive in providing input to managers on safety concerns and other important matters, and that managers and employees understand the value and importance of this input.
8. "Begins with the End in Mind" (Stephen Covey), and "Establishes Continuity of Purpose" (Edwards Deming).
9. Ensures that design, operations, and research and development components of work are fully integrated.
10. Minimizes research in production facilities by production personnel, and production in research facilities by research personnel. The need for innovation and daring by researchers conflicts with the need for discipline and adherence to procedures by operators of production activities.
11. Stays close to its customers to satisfy their needs, anticipate their wants, and provide assurances that quality products or services will be provided reliably and on time. Conversely, ensures that other organizations that provide services or products that are essential to its mission can provide them reliably, safely, and on time.

Strives to obtain products at reasonable costs, but approaches cost-effectiveness carefully. Dr. Deming's sixth point of quality is to "End the practice of awarding business largely on the basis of price." .

12. Champions and frequently rewards excellence, including safety. This helps establish good feeling among employees about their work and helps maintain a good attitude for the importance of excellence, including safety.
13. Ensures that all managers and all employees of the organization are treated with dignity and respect in their day to day dealings. Effective and safe work of the organization is impossible when resentments and hard feelings develop from lack of respect and loss of dignity.
14. Focuses on its management systems rather than its employees, groups of employees, or contractors as the cause of failure, incidents, and accidents. By accepting responsibility for mistakes, management can repair its systems, train employees, develop procedures to prevent failures and accidents, and ensure that the procedures are clear.
15. When adopting complex technology from another organization, ensures that all of the details of that technology are understood and that a significant number of technical and management staff are thoroughly trained.
16. Maintains an accurate record of and learns from its history
17. Selects professional staff with great care, and provides good on the job training and programs for education and self-improvement.
18. Uses its material resources efficiently; avoids accumulation of scrap, excess products, or intermediate products such as spent nuclear fuel; and minimizes production of wastes that require long-term isolation from the biosphere. Some chemical companies have adopted a goal of "no wastes."
19. Maintains an awareness of political, institutional, and global matters that could favorably or adversely impact activities or goals of the organization, maintains a proactive strategy and programs to address these concerns, and ensures that political and institutional leaders are provided full and accurate information about these programs.
20. Accepts the need for strong commitment to important principles of quality, and creates a structure in top management to ensure that the commitments are carried out.

Table 2. Characteristics of Extraordinary Leadership

These characteristics are summarized from an article "The Characteristics of Extraordinary Leadership" in the Winter 1996 issue of <u>The Public Lawyer</u>, by Mr. J. Edwin Dietel, a career government lawyer and extraordinary leader. The article encapsulated the author's American Bar Association publications and review of 171 books on leadership.

1. Trust and Trustworthiness. These are two of the most fundamental characteristics of truly excellent leadership. Few people will be long time followers of a person who is not trustworthy and who they do not personally trust as an individual.

2. Honesty, Integrity, and Setting the Example. Being honest, acting with the highest integrity, and setting the example that we want others to follow are several of the basic or fundamental ways we establish and earn the trust of others.

3. Substantive Expertise and Quality. Few of us will earn the respect that effective leaders need if we lack the substantive expertise that is the subject of the work of those we are attempting to lead.

4. Values Driven. To be effective leaders we need to be driven by some very fundamental values such as fairness, equality, justice, and the several others included in this list of twenty.

5. Motivation. In order to be effective as a leader, a person has to want to lead; the essence of which is being dedicated to helping others. Some simply do not want to be in that position.

6. Passion. With present demands on our time, it is easy to fritter it away being mediocre in what we do. Without passionate commitment to being an excellent leader, it is highly unlikely that one will be an outstanding leader.

7. Dedication and Commitment. The base quality to which the best leaders are dedicated and committed is service; service to others who they want to lead.

8. Patience and Persistence. Best leaders do not get to their positions either by giving up easily or by not being willing to wait for the necessary and appropriate opportunity.

9. Accurate, Timely, and Open Communications. Leadership does not take place in a vacuum. It occurs with others with whom the leader is fully in touch. Communication is considerably more than just one person talking and others listening.

10. "Can Do" Optimism that is Constantly Searching for Opportunities. Leadership is about finding opportunities that others fail to see. It is about the "vision thing" that we will examine below and it is about having an optimistic attitude.

11. Fairness. Some may get away with serving in leadership positions for a while and being unfair, but no leader can survive for long by treating followers this way. Followers rightly expect to be treated fairly.

12. Courage, both Physical and Moral. Courage, or the commitment to fundamental values, is at the bedrock of effective leadership. It is the ability to stand up for and speak out about what the leader thinks is right.

13. Continuous Improvement. While perfection is unattainable, the real leader strives for continual improvement. There is considerable value in working for self-improvement and in motivating others to do the same.

14. Vision. This is the ability to project ourselves into the future and accurately predict what others will want and need in the future. It includes not only the ability to accurately predict this future state, but also to communicate and project this prediction to our followers so collectively we can work to implement a shared vision.

15. Entrepreneurial Driven Spirit. This is a spirit that enables us to be motivated to work independently. It is being a self-starter, and includes an ability to inspire others to want to pursue the vision that we mutually share

16. Being Action Oriented. This is closely related to the spirit just described, but also involves the drive of dissatisfaction. It is the desire to be better and having a will and devotion to do something about it. It is having the spirit to "just do it."

17. Balance and Flexibility. This is not compromise, but a willingness to avoid dangerous extremes. It is a hunger to change for the better, and an ability to accommodate the insecurity that change brings and to marry these often conflicting characteristics of leadership into a package that is most suitable and effective for our own particular personality.

18. Altruism, Humility and Self-sacrifice. These are the concepts of service and selflessness over self and selfishness. They include the willingness and desire to give rather than receive. It is the concept of "we" and "you" versus "I."

19. Mutual Gain. Notwithstanding those ideal concepts of altruism, humility, and self-sacrifice just mentioned, we should mentally drive ourselves to think in terms of mutual instead of just self gain.

20. Respect for People. This is the leader's passionate respect for the humanity of other people and for their ability to contribute to the good of the group and the encouragement and welcoming of them to do that. It is a leader's ability to use the strengths of each person for the collective good of all involved in their particular objective or goal in such a way that they are substantially more than their individual arithmetic sum. And it is being honest, fair, and trusting of others.

President Bill Clinton incorporated Mr. Tobias's ideas of partnerships for quality into Executive Order 12871 for improved operations in the Federal Government. The strong partnership that developed between former Deputy Energy Secretary Bill White and leadership of NTEU Chapter 228 led to partnerships in the Office of Nuclear Energy, Science and Technology, and eventually with Energy Secretary Hazel O'Leary. These partnerships provided an opportunity for good technological input for decision processes, including decisions of Secretary O'Leary to reverse or defer earlier decisions to cancel successful reprocessing programs and decommission the Fast Flux Test Facility.

In April 1997, the Russian Ministry for Atomic Energy and Russian Trade Union for Nuclear Workers agreed to form "Partnerships for Safety and Quality," based on ideas presented to them by leadership of NTEU Chapter 228 and their review of successful nuclear programs in other nations.

SUCCESSFUL USES OF NUCLEAR TECHNOLOGY

Successes with nuclear technology in the United States were achieved by corporations that had:

- good communication systems including accurate information to customers and the public;
- research, development, and design integrated with operations;
- employees empowered for input to managers for safety concerns;
- good safety, environmental, nuclear material safeguards, radiation and fire protection practices and oversight;
- knowledge that problems and solutions were found in the details; and
- good leaders with substantive expertise in the work that they led, and other "Characteristics of Extraordinary Leadership" discussed in Table 2.

Customers with appreciation of the need for quality corporate management and who themselves adopted quality principles and had good leaders were vital for best successes. These included leaders of the Manhattan Project of the Army Corps of Engineers, the Nuclear Navy, some Divisions and Offices of the Atomic Energy Commission (AEC), some nuclear utilities, and more recently, the National Aeronautics and Space Administration.

Examples of success with nuclear technology through use of quality principles include:

- The DuPont Company with the Manhattan Project for safe production of plutonium which was used to end World War II and save hundreds of thousands of Japanese and American lives.
- Energy system vendors with the US Navy and AEC for naval nuclear propulsion systems and with some utilities in the U.S. and other nations for commercial nuclear power plants.
- The DuPont Company with the AEC for production and processing of plutonium-238 for space exploration and tritium for nuclear deterrence, and with the Los Alamos Scientific Laboratory and the Livermore Research Laboratory for quality assured production of reliable tritium weapon components.
- Westinghouse Electric Corporation with AEC reactor development program managers and staff for fuels testing in the Fast Flux Test Facility in support of the U.S. program for development of systems for efficient utilization of nuclear materials which would ultimately eliminate the indefinite and dangerous accumulation of weapon usable fissile and fissionable materials.

- General Electric Company with DOE staff for the U.S. Government nonproliferation initiative with the Government of India, which resulted in success of the initial technical phase and an opportunity for an accommodation between political leaders of the United States and India for enhanced international safeguards of nuclear facilities and materials in India.
- Contractors working with DOE and NASA for production of radioisotope thermoelectric generators that provide electricity for spacecraft. Particularly noteworthy was the full and accurate information provided to the public by NASA and the Jet Propulsion Laboratory that plutonium-238 used for the Cassini mission was safe and essential.

POTENTIAL SUCCESSES REPLACED BY PROGRAMS THAT FAILED

The DuPont Company program for disposition by the AEC of spent fuel from early nuclear power plants would have been an outstanding success if it had continued, not only for recycle of valuable nuclear materials in spent fuel from U.S. reactors with world-best reprocessing technology, but also for elimination of a proliferation threat from supply of U.S. nuclear materials and technology to other nations. Clearances had been obtained from the port authority of every major port in the Eastern United States for import of spent fuel of U.S. origin.

The program was canceled because a commercial reprocessing company relied on an AEC report which indicated that AEC experience in a low cost reprocessing plant had been successful. The company offered commercial reprocessing services based on that low cost. Operation in the low cost AEC plant had not been successful, and that in the low cost commercial plant was not successful.

The program assigned to DuPont by the AEC after failure of commercial reprocessing would also have been an outstanding success, not only for reliable, safe, and cost effective spent fuel disposition, but also because the model developed could have been provided to other nations with strong nuclear programs for maximum assurances against undetected diversions of plutonium.

Plutonium would have been shipped from this facility only in mixed-oxide fuel assemblies. There would have been greatly enhanced accountability for nuclear materials, no accumulation of separated plutonium, and no human access to nuclear materials inside remotely operated and maintained process cells with walls of heavily reinforced concrete up to five feet thick. Readouts of electronic monitors could have been made available to inspectors anywhere in the world. Access to operating corridors would have been only via underground tunnels, through electronic monitors that would have assured no undetected access or removal of materials.

This program was canceled by Energy Research and Development Administration (ERDA) managers who had no understanding of reprocessing plant details and no appreciation of the value of quality corporate management. ERDA fuel cycle programs were revised to carry out research and development in PUREX processing at an autonomous national laboratory, refocused on a different reprocessing plant design concept that had not been successful, and to support another commercial reprocessing plant modeled after the AEC plant that had failed. The different reprocessing plant design concept that had not been successful was that used for demonstration of Experimental Breeder Reactor-II (EBR-II) fuel recycle by pyrochemical processes. The reprocessing plant that had failed was the Idaho Chemical Processing Plant.

If Allied Chemical Company officials had examined ICPP production records, they would have learned that its productivity was equivalent to that estimated by the General Electric Company for its Morris Plant at the time of the GE decision not to operate that facility.

Instead they reviewed reports of company managers at Idaho which implied productivity five times that actually achieved. This led to an assumption that they could profitably build and operate the Barnwell Nuclear Fuel Plant. The low charges offered by Allied Chemical Company forced out of the market corporations that planned to use plant design concepts that were based on successful reprocessing plant operating experience.

AUTONOMOUS NATIONAL LABORATORIES

The X-10 Pilot Reprocessing Plant at Oak Ridge, Tennessee

The facility that led to unsuccessful AEC and commercial reprocessing was the X-10 pilot reprocessing plant, built and operated at Oak Ridge, TN during World War II for process demonstration, prior to production operations at Hanford. Technical staff continued to operate X-10 for recovery of research quantities of plutonium, and convinced themselves and AEC officials that the plant concept was suitable for low cost, production-scale reprocessing.

The ICPP was built on the X-10 concept to reprocess AEC highly enriched uranium spent fuel, including that used for tritium production at the Savannah River Plant. Prior to shut down of tritium production capability because of lack of spent fuel storage capacity from failed reprocessing, the "H-Canyon" PUREX plant was adapted for HEU reprocessing for recycle of the accumulated backlog of spent fuel.

The Experimental Breeder Reactor Fuel Cycle Facility in Idaho

The Fuel Cycle Facility for the Experimental Breeder Reactor II (EBR-II) at the Argonne National Laboratory (ANL) site in Idaho is analogous to the X-10 pilot plant at Oak Ridge. Technical staff at ANL convinced themselves and DOE officials that a modified EBR-II fuel cycle "pyroprocess" was suitable for low cost, "proliferation-resistant" reprocessing. Plans were formulated for a demonstration of this "technology," and EBR-II fuel recycle became the base plan and later the sole plan for U.S. reprocessing.

An examination of details revealed that ANL plans were for a demonstration of process, not technology. Questions of operability, maintainability, safeguardability, and containment of radioactivity would not have been resolved. Aware that the omitted items had been the major problems with commercial reprocessing, DOE reprocessing staff did not support adoption of the EBR-II fuel cycle as the sole plan for U.S. reprocessing.

This examination also revealed great difficulty for material balance measurements and high plutonium losses. These findings, combined with the fact that plutonium is handled in weapon usable form and concentration, led some DOE staff to conclude that the safeguards challenge would be insurmountable. Instead of "proliferation resistant," the process would be a "proliferation threat." Concerns about the IFR fuel cycle are summarized in Table 3.

Some DOE staff also believed that the requirement for inter-process transfer by physical movement of containers of nuclear material instead of through pipes would limit applicability of the EBR-II fuel cycle process to research, or production of small amounts of plutonium. In addition, DOE staff noted that the inter-process containers are not sealed, and concluded that release of significant amounts of contamination into the cell atmosphere was likely.

The pyroprocess equipment complexity; use of exotic materials that are not readily available; combined challenges of a remote environment, inert atmosphere, and high temperature; and other concerns discussed above led DOE reprocessing staff to conclude that the EBR-II fuel cycle process concept of ANL had no potential for use to support commercial nuclear fuel recycle and no potential for low cost or proliferation resistant reprocessing, and should not have been selected as the sole plan for U.S. fuel recycle development.

President Bill Clinton canceled plans for the EBR-II fuel cycle demonstration. The American Nuclear Society Special Panel on Plutonium recommended in 1995 that research continue on this "promising" technology, and a recent report of the President's Advisory Committee on Science and Technology recommended that DOE carry out research on "proliferation-resistant" fuel cycles.

The Los Alamos National Laboratory (LANL) of the Department of Energy has proposed continuing research on "proliferation resistant" pyroprocessing. The LANL approach seems similar to that considered by ANL for the IFR fuel cycle, and will almost certainly have similar problems and concerns.

Table 3. Summary of Concerns about Fuel Cycle Technology for the Integral Fast Reactor

1. Processes to be used for IFR fuel reprocessing were similar to those used for plutonium metal processing in Atomic Energy Commission weapon programs. Much greater difficulties were experienced in plutonium metal processing than in properly designed aqueous reprocessing. Large accumulations of scrap were normal at all plutonium metal plants except for those at SRP where scrap was immediately re-dissolved and returned to reprocessing. In the initial EBR-II fuel cycle experiments, large amounts of scrap were shipped to the Idaho Chemical Processing Plant for recovery.

2. Equipment proposed for IFR fuel reprocessing was much more complex than that used in aqueous reprocessing and would have been very difficult to maintain for reasonable on-stream time under the best of circumstances.

3. Material measurement in the electrorefiner was extremely difficult under cold, development conditions, and was performed only about every year or two in the development facility. Measurement of fully irradiated fuel in a remote environment would be far more difficult, thus material accountability and realistic safeguards would be virtually impossible.

4. High process losses (i.e., several per cent) were experienced, particularly in the fuel fabrication step. There were indications that high process losses would have been likely in electrorefining. This, combined with measurement difficulties, makes significant diversion detection virtually impossible.

5. Operations in a remote environment are about three times as difficult as hands-on operations in glove boxes; operations in an inert environment are similarly more difficult. The combination contemplated for the IFR fuel cycle might be ten times as difficult as those in glove boxes, or about three times as difficult as those in aqueous reprocessing, without consideration of the more complex equipment planned for the IFR process. High temperatures would further increase difficulties.

6. The IFR process requires use of exotic materials that are not available in forms/shapes needed. Research was underway for these materials, but there is no experience base for use of these materials.

7. Inter-process transfer of nuclear materials requires physical movement of containers of nuclear material as opposed to transfer through piping in reprocessing plants that have operated successfully. The containers are not fully sealed, thus there is significant potential for release of contamination into the cell atmosphere.

8. Fissile plutonium is in weapon usable form (metal) and in concentrations usable for a significant nuclear explosive. Some theoretical data suggest that in-process materials may not be directly usable for weapons suitable for military stockpiles, but experimental results with electrorefining suggest that some of the theoretical data needs refinement. Clever operators of electrorefining equipment might be able to produce fairly pure plutonium metal directly usable for military type nuclear explosives.

U.S. NUCLEAR NONPROLIFERATION POLICIES AND PRACTICES

India detonated a nuclear explosive in 1974 using plutonium[2] produced with technology, materials, and facilities supplied by the U.S. and Canada. This action resulted in total withdrawal by Canada from its vital support for India's nuclear power program, and an uncoordinated "triple-track" response by the U.S. that reflects its failure to use quality principles for development of nuclear policies. The three "tracks" were:

- Recommendation to the International Atomic Energy Agency (IAEA) that it conduct a study to determine if "Regional Nuclear Fuel Cycle Centres" would have significant advantages related to the back-end of the nuclear fuel cycle, in addition to making substantial contributions toward the goals of nonproliferation.

- Appointment of a National Security Council Sub-Task Force on Tarapur[3], under leadership of the Department of State, to develop options for disposition of spent fuel of U.S. origin, and possibilities for enhanced international safeguards for India's commercial nuclear program.

- Study by the U.S. Congress Office of Technology Assessment (OTA) "in response to a request from the Senate Committee on Government Operations (now the Committee on Governmental Affairs) to help provide Congress with the capability to "independently evaluate the policy making activities of our Government and other nations and be prepared to take legislative actions" with regard to nuclear proliferation and safeguards."

The IAEA Regional Fuel Cycle Centres Study

An ERDA headquarters reprocessing staff representative provided technical leadership for the "Organization and Administration" and "Spent Fuel Storage" sections of the IAEA study, which concluded that there were advantages to multinational reprocessing from considerations of nonproliferation, waste management, and economics.

In a July 1976 speech to the United Nations Security Council, presidential candidate Jimmy Carter endorsed the multinational reprocessing plant concept and suggested that the partially completed commercial reprocessing plant[4] at Barnwell, SC be used for a demonstration of the concept. President Gerald Ford responded with a White House study leading to a policy statement in late October 1976 that near term reprocessing of spent fuel was not necessary, in effect withdrawing support by the U.S. for the IAEA reprocessing plant study. Knowledgeable reprocessing staff did not participate in the White House study.

---

2 This plutonium was produced in India's Cirus reactor, which was supplied by Canada and was similar to the NRX reactor operated by Canada to produce plutonium for U.S. nuclear weapons programs. Heavy water used as a moderator for the Cirus reactor was supplied by the United States. The plutonium was recovered in the Trombay reprocessing plant, which was designed by a U.S. engineering firm, based on the Oak Ridge X-10 pilot plant concept. India's technical staff was trained at the X-10 plant.

3 The Tarapur Atomic Power Station in India consists of two 210 megawatt (electrical) GE designed light water reactors which were funded by a US AID loan, and nuclear fuel cycle facilities including the PREFRE reprocessing plant, which incorporated the "X-10" design concept. The U.S. had agreed to supply enriched uranium for reactor fuel, under safeguards, and had given consent for India to reprocess the spent fuel.

4 The Barnwell Nuclear Fuel Plant of Allied-General Nuclear Services. Design of this plant was based on the X-10/ICPP concept, which had not been successful in AEC or commercial reprocessing programs.

After taking office, President Carter supported further studies of reprocessing - again without participation by persons with knowledge of the details of reprocessing - and announced plans in April 1977 to indefinitely defer commercial reprocessing and plutonium recycle, and completion of the Clinch River demonstration fast reactor.

The decision to defer reprocessing was made in part to permit an "International Nuclear Fuel Cycle Evaluation" (INFCE) of alternative reprocessing concepts, based on paper studies by persons who were not substantively involved in reprocessing, and had no concept of the important plant operating and design characteristics essential for any serious nuclear proliferation evaluation. The DuPont Company design studies of proliferation-resistant reprocessing, based on its successful reprocessing experiences at Hanford during World War II and the Savannah River Plant from 1950 through 1975, were completed in November 1978, and provided to DOE officials. However, they were not considered in U.S. Government nuclear proliferation studies or suggested for a Regional Nuclear Fuel Cycle Centre.

National Security Council Sub-Task Force on Tarapur

In late summer of 1977, more than two years after start of task force efforts, a DOE headquarters reprocessing staff representative in the Office of Nuclear Energy was invited to a meeting and subsequently requested to visit the Tarapur Atomic Power Station (TAPS) and assess potential for expansion of storage capacity for TAPS spent fuel. Accompanied by a senior manager and a spent fuel storage engineer of the General Electric Company, he met with the U.S. Consul General and India Atomic Energy Commission (IAEC) and India Atomic Power Authority (IAPA) leaders in Bombay. The GE manager had been startup engineer for the two TAPS GE-designed boiling water reactors. The task force was interested in expanding spent fuel storage capacity so that reprocessing could be delayed pending an accommodation on enhanced safeguards for India's domestic nuclear programs and materials.

The assessment indicated that storage capacity expansion was feasible. Collaborative design effort was started by GE under DOE staff leadership and proceeded well, but India's political leaders were opposed to the initiative. No action was taken by India that would facilitate a storage capacity increase until a meeting with IAEC Chairman Homi Sethna in July 1978. The DOE representative pointed out that India's reprocessing plant design was similar to designs that had failed in the United States, and that reliance solely on reprocessing in this facility for spent fuel disposition would almost certainly lead to a shutdown of the reactors and loss of much needed electricity for India. Dr. Sethna's reaction was immediate. Knowing that he could explain the need to political leaders of India, Dr. Sethna ordered senior managers to proceed with the storage capacity increase.

In June 1979, at the request of the Department of State and the DOE Director of Energy Research, the DOE representative was detailed to the DOE Office of Nuclear Non-Proliferation to conduct a study of options for disposition of TAPS spent fuel. The study report was to be background for discussions by U.S. President Carter with India Prime Minister Desai, in an attempt to reach an accommodation on enhanced safeguards for India's peaceful nuclear activities.

A few weeks after start of the study, Prime Minister Desai, with whom President Carter had good rapport, was replaced by Indira Gandhi. The opportunity for an immediate accommodation was lost, but President Carter understood that continuing collaboration with India on peaceful nuclear activities was essential to any accommodation on improved international or multinational safeguards for India's commercial nuclear activity. This would include continued supply by the United States of low enriched uranium (LEU) needed for operation of the Tarapur reactors, which required a license by the U.S. Nuclear Regulatory Commission (NRC). The Nuclear Non-Proliferation Act of 1978 (NNPA) (see next section) specifically precluded supply of this material unless a determination was made by The President that the supply of LEU was in the national security interests of the United States - a determination that, at the time, could be overturned by resolution of both the House of Representatives and The Senate.

The NRC denied a license for LEU supply to India, President Carter made a determination that the supply was in the national security interests of the United States, The Senate upheld the determination, and LEU was supplied for continued operation of the Tarapur reactors. Subsequently, President Carter was replaced by President Ronald Reagan, and a decision was made to withdraw from any further effort on this initiative, consistent with the NNPA.

The OTA Study Report "Nuclear Proliferation and Safeguards"

This report was prepared by the Office of Technology Assessment Energy Program Staff, with the assistance and cooperation of twenty-five contractors and consultants, a "Task Force on Nuclear Weapons," and staff of several government agencies, including national laboratory staff of the U.S. Energy Research and Development Administration (ERDA). Neither OTA staff nor their consultants had reprocessing plant design or operating experience. Experienced ERDA reprocessing staff was not involved in the study nor did it have an opportunity for review until a few weeks before the draft report was approved for publication.

During the last stages of this study, an OTA project staff member contacted an ERDA reprocessing staff representative and asked about a statement in the final draft of the report. The ERDA representative responded that the statement was incorrect. A few days later the OTA staff called back and said that their consultant, a colleague at an ERDA national laboratory, had reconfirmed the statement. The ERDA representative called and resolved the misunderstanding with the consultant, thus establishing credibility with the OTA staff member who then requested detailed review of the entire draft report. The ERDA representative found some 200 inaccurate or misleading statements including many of substance, and scheduled a meeting with OTA staff to review and resolve the concerns. Two days before the scheduled meeting the ERDA Assistant Administrator for Nuclear Energy approved the final draft report for publication.

The OTA report *Nuclear Proliferation and Safeguards* was presented to Senator Edward Kennedy, Chairman, OTA Technology Assessment Board, on June 30, 1977, and by Senator Kennedy to Senator Abraham Ribicoff, Chairman, Senate Committee on Government Affairs. This report provided the basis for the Nuclear Non-Proliferation Act of 1978 (NNPA), which isolated the United States from other nations for best deployment and control of technology for safe, reliable, cost-effective, and well-safeguarded reprocessing, which is essential for viable nuclear power and elimination of the virtually permanent proliferation threat from accumulation and indefinite storage of fissile and fissionable materials in spent fuel.

The nuclear nonproliferation policies of the United States have:

- weakened, not strengthened the international framework for safeguarding plutonium;
- resulted in large accumulations of spent nuclear fuel at commercial nuclear power plants and plutonium at DOE sites;
- led to plans to discard, without use, an energy resource that could supply all of the electricity needs of the United States for hundreds of years, free of greenhouse gases and hydrocarbon and acid pollutants that cause thousands of early deaths each year;
- led to plans to create geologic deposits of large quantities of weapon usable material that becomes increasingly accessible for recovery with decay of short-lived, intensely radioactive fission products, and will be a proliferation threat for millions of years; and
- created a vacuum for good management of nuclear technology in the United States from which recovery will be very difficult.

OTHER FAILURES FROM LACK OF QUALITY CORPORATE MANAGEMENT

- Replacement of integrated GE management at Hanford by AEC management as a result of contractor diversification, which led to changes in policies and many years later in practices for disposition of radioactive wastes, and eventually to decisions to reverse irreversible decisions made earlier on the basis of good technical assessment.

- Transfer by an autonomous AEC National Laboratory of important reprocessing technology which led to nuclear weapon proliferation. Throughout AEC history, important reprocessing technology from production programs was never transferred to other nations, except for classified exchange with the United Kingdom, which did not use the technology.

- Failure of the AEC to ensure that there was good radiation and fire protection oversight at plutonium facilities at Rocky Flats, Colorado, which led to a major fire. Failure of the AEC to accept responsibility for the circumstances that led to the fire, unfair criticism of Dow Chemical Company and technical staff, and replacement of Dow with corporations that had little experience with complex chemical processing operations, has led to many major problems and wasteful expenditures.

- Failure of leaders of the AEC Division of Reactor Development to emphasize the nuclear nonproliferation advantages of fast reactors and to ensure that fuel cycle technology needed to support development and deployment of this technology was based on successful experience. Fast, so-called "breeder" reactors convert separated, accessible plutonium to inaccessible form and ultimately eliminate all potentially weapons-usable fissile and fissionable materials.

MANAGEMENT OF NUCLEAR TECHNOLOGY IN OTHER NATIONS

Other nations, except those with non-market economies, examined successes with nuclear technology in the United States, used or created corporations to carry out or support important nuclear activity, and adopted important principles of quality for these efforts. The United States Government discontinued use of corporations that were experienced in quality management of operations involving complex technology, an action that has led to many problems, and failures.

The Russian Ministry for Atomic Energy and Russian Trade Union for Nuclear Power and Industry Workers have adopted important principles of quality for future nuclear work. As Russia moves forward with democracy and corporate management, these principles will lead to improved safety of nuclear power plants and nuclear facilities, and improved safeguards for nuclear materials.

NUCLEAR WASTE, AND THE LOSS OF CREDIBILITY FOR NUCLEAR TECHNOLOGY IN THE UNITED STATES

During the summer of 1958 I was a participant in a seminar for Chemistry and Chemical Engineering professors at the AEC Hanford Laboratory in Washington State. This seminar was sponsored by the American Society for Engineering Education to introduce courses for nuclear fuel reprocessing and radioactive waste management into curricula in American Colleges and Universities.

My work assignment was with the Environmental Sciences section of the laboratory, conducting chemical and physical studies with simulated radioactive waste to determine behavior of radioactive wastes in Hanford soils.

At that time eight tanks were known to be leaking. In addition, AEC and GE managers at Hanford had been disposing of intermediate and low level liquid radioactive waste into the soil for several years. All of the tests that I participated in or have reviewed since that time supported technical conclusions of GE scientists and managers that there would never be any dangers to human populations or workers as a result of disposal of waste in this manner, or wastes that leaked from the tanks. Any radioactivity that reached the Columbia River from wastes disposed of in Hanford soils would be a very small fraction of a percent of that released to the river for many years as a result of failures of fuel elements in the reactors with "once-through" cooling, or tritium added by natural, atmospheric processes - with no adverse impact on the environment or health of life forms. I did not agree with the decision not to replace or empty leaking waste tanks, but realized that there had been a commitment of Hanford as a permanent repository for radioactive waste several years earlier.

Similar studies by scientists and engineers at the SRP led to similar conclusions, i.e., that the ion exchange characteristics of SRP soils would preclude transport of materials in wastes to drinking water supplies until long after decay of contained radioactivity to levels that would preclude adverse impact to health of any life form. DuPont and AEC managers agreed with the technical conclusions of these studies, but did not adopt the waste management practices recommended by technical staff.

In 1967, the U.S. General Accounting Office assigned teams to conduct in-depth review of radioactive waste management policies and practices at AEC sites. Following completion of the review at SRP, GAO team members discussed findings with the AEC manager and waste management staff. They pointed out two minor incidents where practices had differed from written procedures, but praised the high quality work and safety achievements of AEC and DuPont workers.

The draft report issued by GAO headquarters ignored findings of the GAO team, but instead incorporated input from AEC headquarters staff, who had described certain practices at SRP as dangerous that extensive study by technical personnel at SRP had shown was not dangerous. AEC staff apparently believed that there would be opportunities for career advancement if a separate organization was created in AEC to correct possible dangers. The head of the GAO team that reviewed practices at SRP apologized to the AEC Office manager for the report.

The new Division of Waste Management was created in the AEC, but unlike recent practice in DOE, individuals that had made false allegations of dangers were not selected for senior positions.

Policies at the Hanford site were revised to forbid release of radioactivity to soils, but practices remained the same, because those doing the work knew that the incremental amounts of radioactive waste released to the soil would not add to any potential problem.

In 1986 a senior official of the Washington, D.C. based Environmental Policy Institute (EPI) prepared a report alleging great dangers from radioactive wastes stored at the Savannah River Plant. DuPont prepared a comprehensive report under DOE sponsorship, refuting all allegations of danger.

The August/September 1988 issue of *Technology Review*, publication of the Massachusetts Institute of Technology (MIT), contained the article "Radioactive Waste: Hidden Legacy of the Arms Race" by the EPI official, and an official of the Institute for Environmental and Energy Research in Takoma Park, MD. This article made false claims of great dangers, including a Chornobyl-scale accident from hydrogen in waste tanks at DOE sites.

A similar article "Nuclear Waste: The $100-Billion Mess" by the same authors and with the same false and inflammatory allegations, was the entire first page of the "Outlook" section of the September 4, 1988, issue of *The Washington Post*. *The Post* article included the statement that it was an excerpt from the MIT publication, giving it special credibility.

The DOE had the DuPont report and could have corrected the misinformation but did not. My letter to the editor of both publications with correcting information from the DuPont report and my own experiences was published in *Technology Review*. The Editor-in-Chief admitted at a meeting of the Washington, D.C. MIT Alumni Association that the article was a serious mistake. *The Washington Post* did not publish my letter or other correcting information.

The EPI official joined the staff of the Senate Governmental Affairs Committee in 1989, worked with staff of the Office of Technology Assessment to develop justification for spending hundreds of billions of dollars for "Cleanup of Nuclear Waste," and in 1993 joined the DOE as a Deputy Assistant Secretary for Environmental Restoration and National Security Policy.

Forty billion dollars has been spent by DOE on "Nuclear Waste Cleanup" since publication of the false and misleading information in *Technology Review* and *The Washington Post* in 1988. Many billions of dollars have been spent on radioactive waste management and monitoring at Hanford since 1967. Some of the work has compromised the integrity of the stable waste configuration in old tanks and in the soil; little of value has been accomplished. All of the work has resulted in more radiation exposure and more dangers to humans than if the work were not done. Present estimate in DOE for "Cleanup of nuclear waste" is $400 Billion.

Former Energy Deputy Secretary Bill White started our initial partnership meeting in July 1994 by recalling a thought he had while on the speakers' platform for ceremonies of the completion of a multimillion dollar mill tailings removal project at Grand Junction, Colorado. He asked if the pile was a hazard where it had been, why was it not also at the new site - and if it was not a hazard, why was it moved? I said that if the pile was enclosed, and persons were inside of the enclosure, they would receive large amounts of radiation from radon and its daughters - but if it is not contained, there would be no radiation hazard. I suggested that the justification for removal was jobs for workers, and profits for dirt moving companies.

DISPOSAL OF SPENT FUEL FROM NUCLEAR POWER PLANTS

The American Nuclear Society Meeting to be held in November 1997 in Albuquerque included a topical meeting on "Environmental Aspects of Radioactive Waste Disposal."

My letter to the editor, published in the November issue of *Nuclear News*, says:

"The environmental aspects of radioactive waste disposal are interesting - but what about the virtually permanent nuclear proliferation threat that would result from creating geologic deposits of enough unsafeguarded plutonium to produce 100,000 nuclear weapons?

"Protection from intensely radioactive fission products would disappear after 300 years, i.e., ten half-lives of cesium-137 and strontium-90. After that, the nuclear "waste" could be easily mined, and the plutonium recovered with relatively simple, conventional chemical processing technology.

"I have discussed my concern about creating geologic deposits of plutonium in unreprocessed spent fuel with former Energy Secretary Hazel O'Leary, former Deputy Energy Secretaries Bill White and Charles Curtis, NRC Chairman Shirley Jackson, Senate Energy Committee Chairman Frank Murkowski, former AEC Chairmen Glenn Seaborg and the late Dixy Lee Ray, many colleagues in DOE and the nuclear community, Members of Congress and staff, and many others. Many share my concern; no one has ever presented a credible argument to alleviate this concern.

"Any threat or reasonable concern about nuclear proliferation from nuclear materials in spent fuel could be eliminated by recovering all of the fissile and fissionable materials in well engineered facilities and destroying them in "fast flux" reactors like Hanford's Fast Flux Test Facility (and GE's PRISM). This action would result in the production of electricity equivalent to the needs of the U.S. for the next several hundred years, without the greenhouse gases and other pollutants released from burning of coal, oil and natural gas.

"The problem is that this action would quash the dreams of those who would produce this electricity with windmills, solar cells, and "energy efficiencies" - which could be done at reasonable costs if laws of thermodynamics could be changed or relaxed. Of course they cannot. It would also quash the hopes of those who proclaim that we must "dispose of nuclear wastes" - and the hopes of enemies of nuclear technology who know that we never can."

ENVIRONMENTAL CONCERNS - AND OTHERWISE

In an editorial in the October 13 issue of *The Washington Post*, former Deputy Energy Secretary, CIA Director and MIT Professor John Deutch said "The American public does not now accept the idea that global warming is a potentially significant environmental threat." It does, however, believe that nuclear waste at DOE sites is a dangerous environmental threat.

The American public also believes that plutonium is "one of the most dangerous poisons known," because editions of the CRC *Handbook of Chemistry and Physics* published between 1971 and 1978 said so. It believes that recycle of spent nuclear fuel in the United States is a proliferation threat because public officials have said so - and nuclear program leaders have supported this myth in order to obtain funding for research and development for "proliferation-resistant" fuel cycles.

RESTORING CREDIBILITY OF NUCLEAR TECHNOLOGY

Trust by Americans of nuclear technology was lost because of failure of the nuclear community to ensure use of quality principles. But NASA's recent success in obtaining general acceptance for use of plutonium in space demonstrates that trust for use of nuclear technology can be restored. NASA adopted an essential first principle for quality, "Full and accurate information to the public."

In order to regain and maintain the trust of the American people and obtain full benefit from use of nuclear technology:

- Nuclear program, regulatory, and policy leaders must ensure that quality principles are used by corporations for managing all uses of nuclear technology, and that policies for use of nuclear technology are developed based on similar principles.
- A U.S. corporation should be selected or created to manage nuclear programs that are now assigned to the DOE, including recovery and preparation for irradiation of all fissile and fissionable materials; production of nuclear materials for defense, space exploration, medicine, food irradiation, and other commercial applications (including recovery of valuable nuclear materials in spent fuel); management and long term isolation from the biosphere of <u>all</u> radioactive wastes; and appropriate and cost effective surveillance and environmental management at DOE sites where these programs are carried out. The Nuclear Waste Fund should be made available to these corporations, and present and planned "Environmental Restoration" funds should be made available for appropriate management of the commercial spent fuel for beneficial use, and to correct any dangerous situations, determined through review using quality practices.

- The DOE's best nuclear sites (e.g., Hanford, Savannah River, Yucca Mountain) should be transferred to this corporation; and these sites, our material and financial resources, personnel and technology should be committed to work that will benefit Americans.
- Two or three DOE National Laboratories should be transferred to this corporation. Research at these laboratories should be fully integrated with design and production activities of the corporation.
- The nuclear community, political leaders, and environmental activists should abandon the idea that indefinite accumulation or creation of geologic deposits of weapon usable material is an acceptable practice, and support priority development and demonstration of technology for destroying or rendering inaccessible all fissile and fissionable materials by their recovery and irradiation with fast neutrons for beneficial uses.
- DOE and/or its successor organization should immediately accept responsibility for storage of excess spent fuel in dry cask storage facilities at nuclear power plant sites, begin preparation for delivery to and storage at DOE sites that are appropriate for management of spent fuel, and begin programs to recover and make available for beneficial use the materials in spent fuel, using best U.S. technology, and technology from other nations.
- The nuclear regulatory process should be restructured so that its major functions are to ensure that nuclear programs are managed with quality principles and that there is good safety oversight. The models of fire insurance company oversight for superior risks or corporate oversight for safety and fire protection for self-insured corporations might be adopted for a more cost effective regulatory process that would ensure safer nuclear facility operations.
- The U.S. Nuclear Technology and Energy Policy Board should be established. Good leaders should be appointed as members to develop nuclear and energy policies of the United States based on principles of quality. A clean environment, reliable energy supplies, and protection from threats of nuclear proliferation are as important as our economy, whose policies are developed by extraordinary leadership of the Federal Reserve Board.
- Federal Agencies should ensure that best information is provided to the public on dangers and lack of dangers from use of nuclear technology, that national and international committees and councils formed to set criteria for protection of workers and the public do so on the basis of best scientific information, and that misinformation provided to the public is promptly corrected.

Formidable effort will be required to regain the trust for nuclear technology that existed in the United States during the twenty years following the end of World War II. But it must be done if populations of this planet are to have abundant energy supplies and a clean environment without the dangers of climate and economic disruption and international tensions. Leaders, customers, and sponsors of the nuclear community must lead this effort.

SUSTAINABLE DEVELOPMENT OF ENERGY SUPPLY

Wolfgang Kröger

Paul Scherrer Institut
CH-5232 Villigen PSI

INTRODUCTION

The term "sustainability" or "sustainable development" is not completely new. At the beginning of the 18th century it emerged in forestry and was manifested by instinctive actions of forest management to cope with regional environmental problems (triggering substitution of wood by coal). In our times the "Brundtland-Report" (1987) provided a revival of this idea defining sustainable development as "a development that meets the needs of the present without compromising the ability of future generations to meet their own needs". It contains the concept of needs and the idea of limitations and necessitates a process of change taking environmental, economical, and social aspects into account. The Agenda 21, a program of action as one of the major outcomes of the Rio Earth Summit '92, formally introduces the concept of sustainability development indicators, which are mainly environmentally (climate change) driven.

At the onset of the implementation process the OECD, for example, has been advised to reinterpret Article 1 of its original Convention: its commitment to "sustainable economic growth" (which simply meant sustained annual rises in GNP) should mean growth that sustains human and environmental, as well as economic, capital.[1]

CHARACTERISTICS OF AND THREATS TO SUSTAINABLE DEVELOPMENT

Following the literature on sustainable development numerous definitions can be found, going as far as "sustainability: the only six syllable word for revolution" (Sustainable Seattle newsletters). Two competing interpretations are now emerging: a wider concept with sustainable economic, ecological, and social development, and a more narrowly defined concept largely focused on optimal resources as well as environment management over time. Further, suggested ways to sustainability vary considerably, from fundamental changes to a more pragmatic evolutionary approach (see Table 1).

In any case, sustainable development calls for a worldwide and political strategy. It is also potentially a scientific concept based on the hypothesis that complex systems, such as the supra-system Earth, are organized around the operation of a small number of nested cycles, each driven by a few dominant variables which are presumed to be known and persist over long time periods.[2]

Environment and Nuclear Energy
Edited by Kursunoglu *et al.*, Plenum Press, New York, 1998

In the context of this work sustainable development is regarded as a controllable multi-criteria process towards a broadly defined goal, to be implemented in differential steps on the time axis. For this purpose the more general principles described (fundamental laws, values) need to be made operational by a set of criteria (means of judging) and indicators (measures that quantify and simplify information) which should be robust and allow review of progress.

It is commonly accepted that current patterns of development do not contribute to sustainability; in particular the current energy patterns contribute to unsustainability.[3] The main indications of these tendencies are:

- Accelerated use of non-renewable energy resources, limited role of (new) renewables.
- Doubling of CO_2 and other greenhouse gas emissions within the last 25 years. These are still rising: the values for EU-countries in the year 2000 will probably be higher by 6% compared to 1990; OECD analyses indicate a further doubling of energy related emissions during the next 25 years (without limiting measures).

Table 1. Main directions in the sustainability discussion

Ecological modernisation / eco-social market economy	Structural "ecologisation"
Anthropocentrically marked conviction, that modernization of the existing growth and civilization model is sufficient (innovation).	Fundamental changes in existing structures of thinking, production, and consumption.
"Monetarisation" of ecology: extensive concentration on economical goals, with the idea to solve thereby also ecological problems in parallel.	Equilibrium economy implies a constant ecological capital stock available, based on calculated carrying capacities, from which one tries to derive limiting values for relevant human activities.
Social aspects implicitly represented.	Social aspects are of primary importance; focus on participation as well as equity elements.
Defended principally by - governmental organizations - majority of economists - OECD countries, primarily representatives of the North - World Bank, IMF, BCSD; ultimately also by the WCED	Defended principally by - NGOs - ecologists - non-OECD countries, primarily representatives of the South - some economists and political leaders

- Reduction of the ozone layer, loss of biological variety, adverse effects on cultural land soil and water as natural resources.
- Geopolitical disparities (GNP, level and pattern of energy consumption, distribution of mineral and financial resources as well as welfare, see Fig. 1).

Probably the most overwhelming reason for this challenging situation lies in an unbroken increase of the world's population, especially in large Asian countries currently with small energy consumption per capita, like China, India and Indonesia. Taking this as given and assuming a slight increase in energy consumption per capita in the developing countries we will have to face a worldwide increase in energy demand by 1.5% (WEC) to 2% (Shell) per year, at least over the next 25 years in spite of conservation and more efficient use of energy (see Fig. 2).

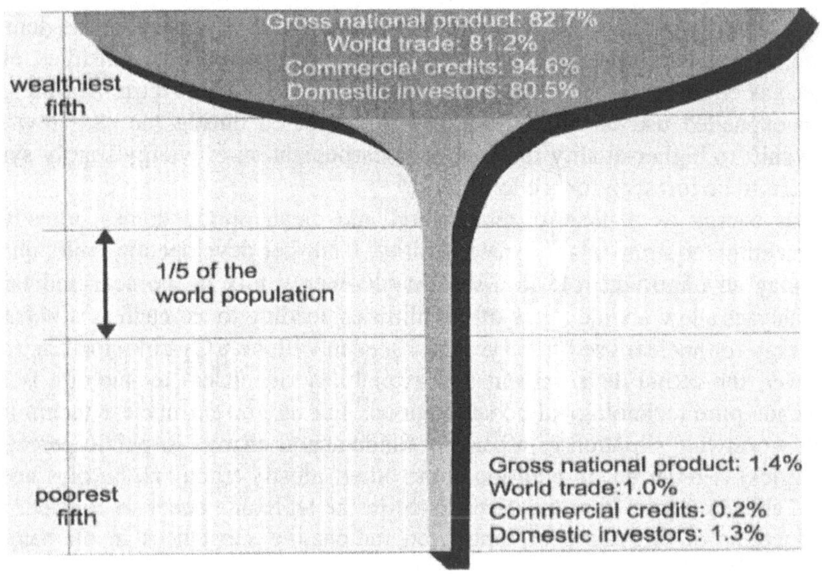

Figure 1. Geopolitical financial disparities

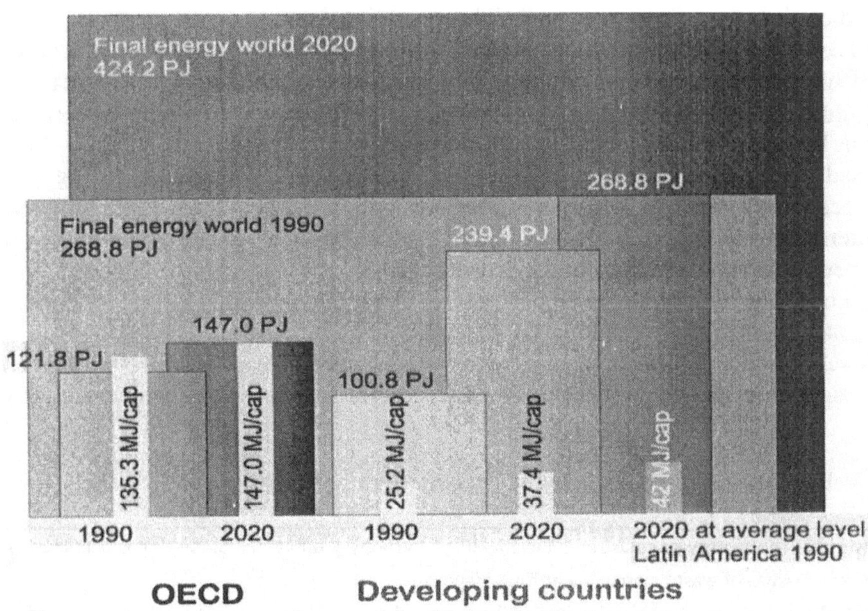

Figure 2. Evolution of final energy demand

TECHNOLOGICAL OPTIONS WITHIN A SUSTAINABLE ENERGY MIX

Following the minimal requirement for sustainable development, i.e. "optimal resource and environmental management", improvements in efficiency on the demand and supply side, as well as progressive reduction of market imperfections, subsidies, etc. in all greenhouse gas emitting sectors, would be the most effective short-term measures. Nevertheless, an expanded use of fossil fuels has to be feared during the next few decades although a shift to higher-quality fuels and a decarbonization of energy supply systems is necessary and to be foreseen for the long-term.[4]

Nuclear energy as a near to zero-carbon and clean option comes, already today, closest to meeting requirements for sustainability. If further developed it could, and probably must, play an important role in a sustainable energy mix in the near and far future. Scenario analyses show a wide range of possibilities, leading to as much as a ¼ fraction of primary energy for nuclear use by the year 2100, even with no CO_2 emission controls.

However, the actual development and availability of nuclear technology is an issue that transcends pure technological considerations. Taking current nuclear technology and the current worldwide situation as a basis, a fundamental change in public perception (in many countries) is required. In addition some other, mostly technical, hurdles need to be taken (see Table 2). Apart from the items listed in the table one needs to consider whether further reductions of emissions of long-lived radioactive substances in all parts of the nuclear fuel cycle are necessary.

REQUIREMENTS FOR A SUSTAINABLE DEVELOPMENT OF ENERGY SUPPLY SYSTEMS

To promote the transition to sustainable development a robust set of criteria and indicators/indices is necessary to make principles operational and to check progress. Such a set has not been established, though been asked for, e.g. by the OECD.[1] The following draft proposal aims at providing input suitable for (comparative) technology assessment and for giving guidance for necessary developments in the energy sector. The information can also be used as basis for trade-off analyses of mixed energy supply scenarios.

Based on the comprehensive definition of sustainable development, taking environmental, economical and social aspects into account, principles have been grouped into three criteria categories:

- "no" degradation of resources in the broadest sense
- "no" production of "non-degradable"waste
- high potential for robustness/long-term stability

whereby "no/non" reflects the aim of being as small/as near to zero as possible. Each criteria category is related to a set of specific criteria and indicators which aim at being

Table 2. Hurdles still to be taken by nuclear energy

- Safety issues (possibility of catastrophic events like Chernobyl), public fears, missing acceptance.
- Unresolved issues of waste management/final disposal.
- Competitiveness issues (presently questionable competitiveness against gas, no readiness for long-term investments, decreasing involvement of industry).
- Limited resources (uranium, thorium) when using a "once-through" fuel cycle; practically unlimited time of use requires reprocessing and fast systems with potential increase of the proliferation risk and public resistance.
- Dependency on a minimum of political, financial and social stability; assured suitability for developing countries.

representative, rather than complete. Whether the current set already is adequate needs to be further discussed; e.g. CO_2 (for global warming) and SO_x (acid rain) are considered as being key parameters for environmental damage while NO_x (photosmog) and small particulates are implicitly represented in the context of health effects associated with normal operation. For allowing specification and objective assessment measures are proposed; for those indicators which are clearly quantitative in nature measuring units are specified.

The assessment should be normalized, for instance based on a plant of 1 GW(e) installed capacity and 1 GWa electrical energy output, taking into account not only the operation of the power plant but the whole energy chain including fuel front- and backend and other elements of the life cycle (LCA approach).

The consensually accepted and suitably filled in matrix (see Fig. 3) could finally be a tool to assess and cross compare current technologies under the constraints of sustainable development, as well as more advanced or even potential options. The weighting and aggregation of the proposed indicators and their subsequent use within decision-making processes need to be developed on the basis of case by case application.

POTENTIAL FULFILLMENT OF PROPOSED REQUIREMENTS BY NUCLEAR ENERGY

Taking current LWR-technology and fuel cycles as basis the strengths of nuclear energy are clearly highly beneficial for environmental protection and climate care ("available clean technology") and security of supply. In order to attempt meeting the whole set of criteria for sustainable development there is a need (and potential)

- to further improve the safety of nuclear power plants to such an extent that catastrophic events can be (entirely) excluded and
- to optimize fuel technology and cycles in terms of "no" net resource consumption, proliferation resistant reprocessing and recycling steps, "burning" of extremely long-lived actinides and fission products (thus reducing guaranteed confinement times in final repositories to a "historical" scale).

With regard to safety recent studies demonstrate that well-operated plants of western design carry small risks compared to other energy supply systems (see the contribution of S. Hirschberg to these proceedings). However, catastrophically high consequences in terms of latent fatalities and land contamination have been estimated although the associated frequencies are almost negligible.

For advanced (next generation) systems under development the design basis has been broadened substantially, such that in-plant measures against severe accidents have been taken into account rendering emergency actions outside the plant unnecessary from a technical point of view. In parallel developments are underway aiming at establishing a new, even more convincing safety quality through radical changes in reactor design. Improved inherent or passive safety features such as easily manageable systems for after-heat removal and fission-product retention, should replace complex active systems and guarantee less dependence on correct operator actions. To assure a higher extent of "error forgivingness", water supplies are designed to keep the reactor in safe conditions even without additional feed from the outside during the first few days after a hypothetical potential accident. Integral testing of passive systems for heat removal from containment under LOCA conditions for simplified BWR (GE-SBWR/ESBWR, Siemens SWR-1000) has demonstrated their functionality and robustness.[5]

Furthermore, higher competitiveness should be achieved by means of simplification of concepts, systems and licensing procedures, standardization and certification, optimi-

Criteria categories	Specific criteria	Indicator		Measure	Current	Advanced	Potential
"no" degradation of resources	Use of fuel and materials	Depletion time[1)]		years			
	Land losses	Non-usable area × duration	Operation	km^2 × years*			
			Severe accidents[2)]	km^2 × years			
	Environmental impact	GWP		tons CO_2 equiv.*			
		Acid rain		tons SO_x equiv.*			
	Human health effects	Acute and latent fatalities	Operation	Number of fatalities*			
			Severe accidents[2)]	Number of fatalities			
			Obj. risk estimates	Number of fatalities*			
	Social aspects	Acceptance		Δ%			
		Proliferation risk		qualitative			
		Economic competitiveness		Currency Unit/kWh			
"no" production of non-degradable waste	Amount of waste			m^3 or tons*			
	Confinement time[3)]			years			
high potential for robustness, long-term stability	Supply security	Foreign dependency		qualitative			
		Technology availability[4)]		Currency Unit			
	Failure friendliness	Grace periods[5)]		hours			
		Socio-political conditions		qualitative			

(Column group heading: Technologies)

* per GWa

[1)] assuming stabilisation at today's generation level
[2)] maximum value identified by risk analysis for a 1 GW(e)-plant
[3)] necessary to achieve "natural levels"
[4)] expressed in expected costs of R&D effort invested until commercialization
[5)] period of time after an abnormal event before human corrective actions become necessary

Figure 3. Proposed criteria and indicators for technology assessment (covering the full energy chain) considering principles of sustainable development

zation of the fuel cycle (e.g. higher burn-up), extended life-time and preventive maintenance.

With regard to fuel technology and fuel cycles more efficient use of uranium, and later also of thorium, should be the primary long-term goal for sustainable development with respect to resources. In the short-term this can be achieved by higher burn-up and by increasing the use of plutonium as energy source.

Proliferation risks are judged differently in the open literature. Some experts do see them as one of the key issues.[4] Others refer to "no-case-of-... experience" over the last decades and turn it down as a misperceived artificial problem. As a matter of fact the results of the INFCE-study (1980) indicate that it is "not possible to make the nuclear fuel cycle absolutely "proliferation-proof" by technical fixes only" but obviously "integration of reprocessing in the reactor plant is advantageous" and "increased build-up of higher isotopes through multiple recycling makes plutonium less attractive for theft". Resolutions of the more recent Moscow Summit Conference and follow-up meetings in 1996 claim safe handling of fissile material as being imperative and give advice on suitable technical and organisational measures:

- Re-inforcement of the IAEA Safeguards System and export-control measures.
- Dilution of weapons-HEU into LEU for LWR fuel.
- Use of weapons-Pu as MOX-fuel in (existing) reactors.
- Use of MOX fabrication plants, construction of an additional one in Russia.

There is little doubt that, in the current situation of world-wide growing plutonium stockpiles (in irradiated reactor fuel and from dismantled warheads), the regulation of plutonium-inventories is an important issue from the non-proliferation viewpoint. As indicated above, the use of MOX fuel in LWR's (ranging from the current-day maximal 1/3-core loadings to 100% MOX cores in the future) is a possible solution. Potentially, uranium-free inert matrix fuels (IMFs) could be deployed in this context, and fast reactor systems offer additional opportunities (see Table 3).

Table 3. Current and future fuel cycles for the regulation of plutonium inventories

(A) Net plutonium production of LWRs and fast reactors

Reactor type	Fuel	kg/GW(e)-a
LWR, without recycling	UO_2 enr.	250
LWR, "self sufficient"	UO_2 enr./MOX	≈0
LWR, 100 % MOX	MOX	-500
LWR, U-free (IMF)	PuO_2	-1260
FBR, breeder	UO_2 / MOX	250
FBR, reduced blanket	UO_2 / MOX	≈0
CAPRA*, MOX	MOX	-600
CAPRA, U-free (IMF)	PuO_2	-1050

* Pu-burning fast reactor

(B) Use of the energy content of uranium:

- LWR without recycling: 0.5 - 0.6%
- Pu-recycling in LWR: 1.1 %
- fast "closed" systems: about 80 % (limited by reprocessing losses)

The development of IMFs, e.g. a solid solution of Pu, Zr and Er oxides for LWRs,[6] would allow efficient incineration of plutonium of civilian and military origin in the short- or at least mid-term. The advantages are as follows:

- No additional plutonium build-up
- Increased proliferation resistance (higher build-up of fission products, insolubility in acids).
- Reduced migration from the final repository (insolubility in ground-water).

In the long-term it is, of course, the deployment of fast reactors as Pu-breeders which can extend the depletion time of uranium by a factor of about 150, i.e. almost to infinity (see part B of Table 3). Such closure of the fuel cycle would also reduce the radiotoxicity of wastes by an order of magnitude.

To further narrow requirements for the final repository and to reduce secured enclosure times to a historical scale partitioning and transmutation (P&T) of minor actinides (Np and Am, in particular) and even of long-lived fission products would be essential. This would require dedicated fast and accelerator-driven (hybrid) systems, in symbiosis with LWRs and FBRs. Studies and developments, mainly in France (SPIN) and Japan (OMEGA), are underway, and Table 4 gives an overview.

CONCLUDING REMARKS

Although sustainability is increasingly regarded as the key paradigm for global development a clear, commonly accepted unique definition does not, and perhaps cannot, exist. A set of criteria and indicators making underlying principles operational needs to be developed. The present proposal aims at providing input in the context of energy systems. It has been applied to current, advanced and potential nuclear systems/options. The results (see Fig. 4) are qualitative and preliminary. Nevertheless, they clearly indicate basic drawbacks and merits and could give guidance with respect to technology assessment and specific developments.

Nuclear energy has the potential to play a major role in a future, more sustainable energy mix. Along the way it has to overcome difficult hurdles which clearly transcend purely technical considerations.

Table 4. Actinide transmutation (complete actinide fission)

Integrated system:

- Transition from a LWR- to a pure FBR-strategy
- Retention of all actinides in the FBR fuel cycle. Reference: IFR (ANL); metallic fuel with electrochemical reprocessing.

Double-Strata Fuel Cycle:

- Symbiosis LWR/FBR.
- Separation of Np, Am, Cm and burning in hybrid system with fast neutron spectrum.
- 1 transmutation reactor supports about 15 normal reactors. Reference: Molten Salt Reactor (JAERI).

LWR with nuclear phase-out:

- Only LWR without or with limited Pu-recycling.
- Burning of the remaining minor actinides in a hybrid system. Reference: ATW (LANL).

Criteria categories	Specific criteria	Indicator		Measure	Current LWR; "once through" fuel cycle	Advanced LWR; "new" safety quality, partly closed fuel cycle	Advanced LWR and FR, dedicated systems; closed fuel cycle
"no" degradation of resources	Use of fuel and materials	Depletion time[1]		years	–	o	+ (+)
	Land losses	Non-usable area × duration	Operation	km^2 × years*	+	+ +	+ +
			Severe accidents[2]	km^2 × years	– –	o	o
	Environmental impact	GWP		tons CO_2 equiv.*	+ +	+ +	+ +
		Acid rain		tons SO_x equiv.*	+ +	+ +	+ +
	Human health effects	Acute and latent fatalities	Operation	Number of fatalities*	+ +	+ +	+ +
			Severe accidents[2]	Number of fatalities	– –	o	o
			Obj. risk estimates	Number of fatalities*	+	+ +	+ +
	Social aspects	Acceptance		Δ%	– –	–	o
		Proliferation risk		qualitative	o	o (–)	?
		Economic competitiveness		Currency Unit/kWh	o	o	?
"no" production of non-degradable waste	Amount of waste			m^3 or tons*	+ +	+ +	+ +
	Confinement time[3]			years	– –	–	+
high potential for robustness, long-term stability	Supply security	Foreign dependency		qualitative	+ +	+ +	+ +
		Technology availability[4]		Currency Unit	+ +	+	o
	Failure friendliness	Grace periods[5]		hours	– –	+	+
		Socio-political conditions		qualitative	–	– –	– –

* per GWa

[1] assuming stabilisation at today's generation level
[2] maximum value identified by risk analysis for a 1 GW(e)-plant
[3] necessary to achieve "natural levels"
[4] expressed in expected costs of R&D effort invested until commercialization
[5] period of time after an abnormal event before human corrective actions become necessary

Figure 4. First, qualitative application of proposed criteria and indicators for technology assessment to nuclear energy chains considering principles of sustainable development

ACKNOWLEDGMENTS

This paper has greatly benefited from advice, information, and help provided by S. Hirschberg, R. Chawla and K. Foskolos, all of PSI.

REFERENCES

1. High-Level Advisory Group on the Environment. Report to the Secretary General of the OECD, (1997).
2. S. Murcott, (MIT). Sustainable Development: A Meta Review, paper presented at Annual Conference of AAAS, Seattle, (1997).
3. WEC, Report No. 7. Climate Change Negotiations: On Track for Kyoto (1997).
4. N. Nakicenovic, Long-term perspectives: energy, development and the environment, *21st Annual Symp*. Uranium Institute, London (1996).
5. J. Dreier, M. Huggenberger, C. Aubert, Th. Bandurski, O. Fischer J. Healzer, S. Lomperski, H.J. Strassberger, G. Varadi, G. Yadigaroglu, The PANDA facility and first results. *Kerntechnik 61, 214*, (1996).
6. G. Ledergerber, C. Degueldre, U. Kasemeyer, A. Stanculescu, J.M. Paratte, and R. Chawla. Using Civilian Plutonium in LWRs with an Inert Matrix Fuel (IMF). *In Proceedings of the International Conference on Future Nuclear Systems,* Yokohama, (1997).

THE IMPORTANCE OF NUCLEAR ENERGY TO SUSTAINABLE DEVELOPMENT

Jean Pierre Rougeau

COGEMA

INTRODUCTION

In December 1997, the UN Framework Convention on Climate Change will hold its third session in Kyoto, Japan (COP3). This conference is an important step in the international negotiations initiated at the Earth Summit (Rio de Janeiro), 1992, where the international community first recognised the need for urgent action to combat the Greenhouse Effect.

To mark the 5th anniversary of the Rio Conference, the UN General Assembly held a special session attended by the heads of government in June this year. The same concerns were reiterated; greenhouse gas emissions had to be curbed if profound climatic changes were to be avoided. Despite substantial agreement on this aspect, there was no general consensus on how the necessary reductions in carbon dioxide emissions could be achieved.

Less than two months to go before the COP3, it is worth noting the developments so far:

- the European Union has put forward a clearly defined strategy, proposing a 15% reduction in 1990-level greenhouse gas emissions by 2010.
- Japan has proposed that industrial nations cut their emissions of greenhouse gases by 5% from their 1990 levels between 2008-2012.
- the US is ready to reduce emissions to 1990 levels between 2008-2012 on condition that other countries do the same.

In sharp contrast to the previous conference on climate change, the potential of nuclear energy for limiting greenhouse gas emissions is now recognised. This is due not only to several nuclear organisations (FORATOM and The Uranium Institute, in particular) but also to other international business organisations (World Business Council for Sustainable Development, International Chamber of Commerce) drawing the attention

Environment and Nuclear Energy
Edited by Kursunoglu *et al.*, Plenum Press, New York, 1998

of negotiators to the contribution of nuclear power today and to its future potential as an energy source in a sustainable world.

Nuclear power has an invaluable role to play in a sustainable future. It is a clean energy source, capable of providing substantial quantities of electricity at a competitive and stable cost and the small volumes of waste it does produce can be well isolated from the environment.

NUCLEAR POWER PRODUCES ELECTRICITY WITHOUT THE EMISSION OF GREENHOUSE GASES

In 1992, global warming had become a priority issue and tangible measures were urgently needed. At the Rio conference targets were set for the industrialised countries to return their greenhouse gas emissions to 1990 levels by the year 2000. Despite these good intentions, carbon dioxide levels (as a result of burning fossil fuels) have risen by 4% in the OECD and 25% in developing countries between 1990-1995 and as yet a potential global solution is not in evidence. If no action is taken to combat global warming, the WEC projects CO_2 emissions in 2020 to be approximately 40% above 1990 levels as the demand for energy increases. Global electricity demand alone is expected to increase by 100% by the year 2020, closely following trends in economic growth and changes in population.

If sustainability is a serious goal, this problem has to be addressed now and not in 10 or 20 years time. Today's investments will dictate how this electricity is generated in the next forty years which in turn determines the extent to which the environment will be adversely affected.

Partial solutions are numerous; developing renewable energy sources and improving energy efficiency are both desirable and will assuredly have an important role to play in the future energy scenario. However, there remains a large unfilled gap between the forecast energy demand and the amount of energy which can be produced or saved by these sources.

- The WEC predicts that renewable energy could increase to 5-8% of total energy supply by 2020 in a 'supportive' environment. This seems rather optimistic as renewables such as geothermal, solar and wind power and biomass presently produce less than 1% of electricity in the OECD countries. Of course, this may change as technology advances - it is likely that renewable technologies will indeed have a key role to play in the future energy balance, but it could be well into the next century before they are developed for extensive commercial application.
- The second proposed option - energy savings - will also improve as innovation is encouraged and existing technologies are developed. But, there is no feasible way in which they can compensate for the increase in energy consumption of the anticipated order of magnitude. To give an idea of the extent of the improvements required, the situation can be quantified as follows: if total CO_2 emissions are to be stabilised by the year 2020, energy efficiency must be increased from 0.42toe/k$ 85 in 1990 to 0.2toe/k$ 85 in 2020 on a global level. This is a very ambitious target and may not be realistic.

Nuclear power currently produces 17% of the world's electricity (7% of primary energy) with more than 440 reactors in 32 countries. This avoids 2.3 billion tons of CO_2 emissions <u>every year</u>[1] - approximately 10% of total emissions. In Europe alone, nuclear

1 Assuming it would otherwise be provided by coal fired plants

power reduces CO_2 emissions by 700Mt per year which is the equivalent of taking all cars in Europe out of circulation. A 1000MW coal power station on the other hand, produces approximately 6.5Mt CO2 in a single year and a gas power station just under half this amount. This is in addition to large quantities of other greenhouse gases such as NO_x and SO_2. A nuclear power plant does not produce any of these greenhouse gases.

It is estimated[2] that the present capacity for nuclear energy could realistically be quadrupled by the year 2050 (from 340 GWe to 1200GWe or from 440 to 1000 reactors world-wide), simultaneously reducing world-wide CO_2 emissions and bringing us one step further towards a sustainable future. The Forum Scenario developed by the European Commission (DGXVII) in its report "European Energy to 2020[3]" provides a comprehensive and detailed view of how coherent energy policies, adopted on an international basis, can affect environmental and economic trends. An interventionist approach is adopted where energy policy is driven by global environmental concerns and sustainable development rather than purely economic criteria. Admittedly, the scenario concentrates on the European rather than the global situation, but it demonstrates how consensus and co-operation between governments and institutions, can in fact produce the desired effects and substantially cut CO_2 emissions. With nuclear and renewable energy sources providing a third of energy needs, this scenario shows that carbon dioxide emissions in Europe can be reduced in 2020 to 89% of 1990 levels.

A climate conducive to energy investment, encouraging energy efficiency and conservation across the energy sector and society as a whole, will be an important part of a future response but not the ultimate solution. The answer lies in finding the optimum combination of energy intensity, nuclear energy and renewable energies. They all have complementary roles to play and we should keep the options open.

NUCLEAR WASTES ARE MANAGEABLE

The disposal of radioactive waste is an important issue in the nuclear sector and is often used to call into question the sustainability of nuclear energy because of the legacy it leaves for future generations. Yet, unlike other industries, the nuclear industry does not release harmful gases indiscriminately into the atmosphere or dump toxic wastes in an uncontrolled manner. For example, a 1000MW coal power station in addition to its CO_2 emissions, also produces approximately 400t/year of heavy metals (e.g. lead, cadmium, mercury) and other toxic chemicals.

Only a small amount of nuclear waste is produced annually which is easily identified and detected, and can be sorted into well defined categories of waste which then undergo the appropriate treatments. The waste becomes less toxic with time and due to the very limited volumes involved and the substantial investments that have been made on both a national and international scale, it is now technically and economically possible to isolate it from the environment.

From the very beginning, the nuclear industry has faced up to its responsibilities towards society, the environment and future generations; it has adopted a waste management policy which meets the highest safety requirements. Reducing the volume of

2 The Uranium Institute, Responding to Global Climate Change: The Potential Contribution of Nuclear Power 1997
3 European Commission : Energy in Europe, Spring 1996

programme. An example of this is the RCR option (reprocessing-conditioning-recycling) which allows both the volume of waste and the final radiation level of the waste to be reduced (the closed cycle produces $0.5m^3$/tonU of waste).

Radioactive waste is obviously a sensitive issue but solutions do exist provided a long term strategy (e.g. RCR) is adopted. The problem is in fact political and not technical. The concerns of the public must be addressed and the case for nuclear power put forward.

NUCLEAR POWER IS WELL ADAPTED TO SUPPLYING CONCENTRATED INDUSTRIAL CENTRES/URBAN DEVELOPEMENTS

Fuel costs for nuclear energy are a very small component of overall costs. It can be seen that total cost per kWh is very similar for gas and nuclear however the composition of this cost differs quite significantly. Investment costs per kWh of electricity generated by fission are high compared to investment costs for a kWh generated by natural gas. On the other hand, fuel costs are less significant and a doubling of fuel price hardly affects the overall cost of a nuclear kWh but practically doubles the cost of a gas kWh.

If nuclear power has lost its edge slightly to electricity generated by combined cycle power stations it can be overcome by advances in nuclear fuel design and management techniques and standardisation of reactor designs. In addition, for the combined cycle power station to be an economically viable option, the demand for electricity must be accompanied by a simultaneous demand for the heat produced and this is not always the case.

Due to this small variable cost, nuclear power stations are cheap to operate and particularly well suited to stable, base load operation. Heavy industrial or densely populated areas are ideal customers as there is constant energy demand and the nuclear electricity utilities can therefore offer competitive and stable rates, particularly important for energy-intensive industrial users competing on international markets.

Renewable energy sources are presented with several disadvantages when faced with providing large quantities of energy. The intermittent nature of wind and solar energy makes it difficult to provide large, reliable quantities of energy at competitive costs. The space needed to generate this type of energy can also place constraints on choice of site e.g. a peak 1000MWe power station would need 25-50 square km if solar powered or 50 - 150 square km if wind powered. In addition, load factors for both solar and wind, under average European climate conditions, are 25% at best.

As sustainable development gains momentum, the price of energy is going to include an internalisation of environmental and social costs which have until now been downplayed by all but the nuclear industry. The electricity utilities will be required to incorporate this extra cost when using fossil fuels. This will enhance nuclear power's competitiveness in the face of deregulation and an increasingly competitive market place.

URANIUM RESSOURCES ARE ABUNDANT AND DIVERSIFIED

Fossil fuels presently provide 64% of world electricity but this will have to change. Even if the lifetime of gas and oil fields can be extended through improvements in technology, the resources are limited. Fossil fuels are not only a source of energy; they have alternative uses which are often of greater added value and for which they are irreplaceable e.g. petrochemicals, plastics, not to mention transportation... It would therefore make sense if they were reserved solely for these purposes, especially in industrialised countries which can access other sources of energy.

According to the OECD/NEA, known uranium reserves represent 90 years of consumption at present levels though this can be easily increased if other sources are taken into consideration such as phosphate deposits or sea water which both contain uranium at low concentrations.

Uranium resources can also be extended by recycling uranium and plutonium. Alternatively, plutonium recovered from dismantled nuclear weapons can be used in MOX fuel, taking weapon-grade plutonium out of circulation and saving valuable uranium resources. (150 tons of weapon-grade plutonium can provide 1000TWh corresponding to half the electricity currently produced by nuclear power in the world in a year). In fast neutron reactors, plutonium is produced as a by-product of electricity generation ; this can further increase resources by up to 2500 years at current consumption levels.

Inherent reliability of physical energy supply is vital in view of the fundamental role energy plays in the functioning of a nation's economy - reliance on another country for fuel puts the recipient in an extremely vulnerable position. Uranium reserves are however well distributed throughout the world, with large ore deposits in what are generally regarded as politically stable countries e.g. Australia (27%), Canada (13%), US(5%), and to some extent Africa (9%). Substantial reserves also exist in the CIS. This distribution of resources allows for diversity in fuel suppliers and avoids potential conflicts resulting from the geographical concentration of oil reserves. Nuclear fuel is also easily stocked and several years supply can be stored without difficulty. In comparison, oil stocks are of the order of only several months as the volumes involved are much larger.

The electricity generation system can undoubtedly rely on a sustainable nuclear fuel supply.

CONCLUSION

Sustainable development is more than just a climate change issue. Future energy sources must satisfy many additional criteria, such as economic viability, sustainability of resources and security of supply.

A pragmatic approach must be applied which will entail a serious examination of public concerns to determine whether or not they are justified. The public should be given a full, impartial assessment of the benefits, costs and risks involved for all the options available, including nuclear power.

The nuclear industry has the opportunity to proactively position itself as an environmentally-friendly, competitive and secure energy source. Its objectives are as follows:

❶ To increase flexibility and cost effectiveness in all aspects of reactor design and fuel management.

❷ To recycle as much as possible to preserve natural uranium resources.

❸ To continue to develop processing and waste disposal technologies with the aim of decreasing volumes and radio-toxicity levels of waste produced (i.e. the principle known as ALARA -as low as reasonably achievable). This will simultaneously facilitate storage and help allay public apprehensions.

Increased research and development for nuclear energy was recently supported by the President's Committee of Advisors on Science and Technology. In a report summary published in September 1997, this committee recommends increasing three-fold US R&D investment into nuclear fission systems, considering the potential benefit from nuclear power in alleviating the problems posed by carbon dioxide emissions, as warranting such an initiative.

The future is going to be a challenging one for the nuclear sector but the opportunities exist. Whether it is able to take full advantage of them will depend on the flexibility of its approach, on the extent to which it actively engages in the process of gaining public acceptance and also on policy makers perceiving clearly the challenge.

REFERENCES

1. WEC, 1995.
2. AIE, 1997.
3. Energy in Europe, DG XVII, European Commission 1996
4. BP Statistical Review 1996
5. OCDE/AIEA 1996
6. Federal Energy Research and Development for the Challenges of the Twenty-first Century (President's Committee Advisers on Science and Technology), 1997.

NUCLEAR POWER AS A BUSINESS: FINANCING THE NUCLEAR RENAISSANCE

Shelby T. Brewer, President

S. Brewer Enterprises, Inc.
29 Blue Ridge Drive
Simsbury, CT 06089

I am sure that it is clear to most participants at this conference that the world cannot do without new nuclear power deployment over the next fifty years given:

1. environmental constraints;
2. the ultimate depletion of economically recoverable fossil resources;
3. demand growth, even at modest levels, in industrialized nations; and
4. demand in the developing world for industrialization and standards of living approaching those of industrial nations.

The energy balance simply does not close without a significant further contribution of nuclear power in the energy mix[1].

Despite the fact that nuclear power is an imperative, nuclear power is stalled in most of the world. Asia, of course, is the exception. In the U.S. the reasons for the downturn are well understood: a planning and construction overhang in the early 1970s; lack of plant standardization, and the resulting micro-regulation by the NRC; loss of political and public support; a stagnant economy in the in the late 1970s, featuring high inflation and interest rates, which aggravated the impacts of protracted construction schedules; and acute state rate regulatory pressures. Many nuclear utilities and suppliers absorbed ruinous losses, and in the end (late 1970s), the financial community lost confidence in nuclear investment, except for supplying services and fuel to operating nuclear plants.

The underlying reasons for the downturn were all institutional/political/economic, and do not reflect an inherent fault of the technology. Therefore, it is innovation on the financial and institutional fronts that is relevant -- not the wistful contemplation of new, "acceptable" technology to be developed at the taxpayers' expense. The notion that new nuclear reactor species are necessary to regain public acceptance is nonsense. To the public, reactor A is indistinguishable from "safer" version B, and no amount of techno-jabber will make a dent in

[1] Yet the current U.S.Administration, in announcing its world environmental goals last week, failed to render nuclear energy even a footnote. This is remarkable in that nuclear power is the most environmentally benign power generation method, and one of the few options available to abate further global warming. This week, however, we expect the Administration to announce a policy allowing nuclear technology exports to Mainland China. Far be it for me to look a gift-horse in the mouth, but this event has more to do with finding a sound-byte for the week than it does with initiating a fundamental new direction, rationale, and commitment. The foundation for this policy was laid painstakingly and carefully years ago by Ambassador William Kennedy, during the Reagan/Bush Administrations.

Environment and Nuclear Energy
Edited by Kursunoglu *et al.*, Plenum Press, New York, 1998

deeply imbedded lay-phobias. Further, the notion invalidates the sound argument that we have been making all along: that the current generation of U.S. reactor designs is "safe enough", indeed less hazardous to humans and the environment than other forms of power generation. We must quit caressing our institutionally-dead bride, and apply ourselves to restructuring nuclear power as a *business,* rather than perpetuating technological sandboxes with federal largesse. Nuclear power will return, not by deploying new "safer" technology, but rather when there is a market "pull", and when the financial community perceives this pull, and has confidence in regulatory and policy stability, and in the competence of operator-owners to manage their fiduciary responsibilities.

One of the curses of U.S. nuclear power is that it has been a ward and an orphan of the federal government – locked in a fatal embrace with Uncle Sam. The U.S. government has been a nag and a nanny to a seemingly sick and dependent industry. The U.S. government during the Carter and Clinton administrations has been a notoriously unstable force in nuclear power: the fanciful notion that the USG should intervene to curtail nuclear power, or at best, regard it as an option of last resort; cooperation projects, such as Clinch River, interdicted after significant private investment; failure of the government to perform on its basic statutory obligations, like providing by a date certain a national high-level waste system; recent abrogation of contracts, such as the uranium enrichment contracts signed with utilities in the early 1980s; willful fiscal mismanagement of the nuclear waste fund legislated in 1982; and I could go on. What is vitally needed is a large, and perhaps painful, dose of Adam Smith in the nuclear sector: get Uncle Sam out of the business except for a few fundamental statutory obligations which it has failed on to date, and place these residual obligations under new management.

Financing and utility sector restructuring are the key factors in any resurgence of nuclear power in the U.S. In my last paper before this conference[2] (November 1995) I presented some alternative commercial structural models in which nuclear might return as a healthy industry:

1. Nuclear Monopoly
2. Status Quo
3. Turnkey
4. Vendor Equity
5. Nuclear Independent Power Producers (NIPPs).

At that time, I rejected both ends of this spectrum, i.e. (1) Nuclear Monopoly and (5) NIPPs, but now, given the Northeast Utilities (NU) calamity in the meantime, I have changed my mind. The NU problem, and the inept NU management response, have taken U.S. nuclear utility sector back into a sink-hole, after years of industry progress in re-gaining regulatory confidence in the industry and major strides in streamlining the regulatory process. The NU episode gives the entire industry a black eye with the NRC, with the state rate regulators, with ratepayers, with the public, and with the financial community. NU is not an isolated example. Aside from a handful of well managed utility companies, the challenge of managing diverse and complicated responsibilities – transmission and distribution, power generation, construction and maintenance, capital formation, wheeling, wholesale purchasing and selling of power, cost-containment, dealing with economic and environmental regulatory bodies, dabbling in real estate and other non-related sectors through non-regulated subsidiaries, *and managing nuclear generation*--- is simply too great of a challenge. Therefore I believe that incremental approaches to a new commercial format – (3) turnkey and (4) Vendor Equity – will not work, and that the most propitious course is the most radical – NIPPs.

I envision a few large NIPPs operating nationally in a competitive environment, selling wholesale and retail power. The NIPPS would be consortia of NSSS suppliers, architect-

---

[2] Shelby T. Brewer, *Nuclear Power in Its Second Half Century*, Remarks before International Conference on Economics and Politics of Energy, November 27, 1995, Plenum Publishing

engineering firms, with the critical mass of expertise and financial wherewithal to build, own, and operate nuclear power plants, and to acquire and operate existing plants. The NIPPs would focus sharply on profitable, competitive, safe nuclear generation and nothing more – unlike the plethora of diverse responsibilities faced by traditional integrated utilities. Integrated utilities, with generating and transmission and distribution functions will probably become a thing of the past.

The 1992 Energy Policy Act offers opportunities in this regard. The 1992 Act enabled, wholesale deregulation, and paved the way for retail deregulation and restructuring as well. Well-managed utility companies are learning how to prosper in the new competitive environment; poorly managed ones are not. Let me emphasize this point: if you want to participate in a nuclear renaissance, then understand and *lead* the retail restructuring debate now unfolding in the U.S. Congress and state legislatures.

The mega-trend in this industry, which will profoundly determine nuclear power's fate, is the inexorable march toward deregulation of *power generation* functions on a wholesale and retail basis. That means that a homeowner or commercial entity or a factory can choose his own power generation company, and that ultimately the only rate-regulated piece in the chain will be transmission and distribution. It means that price and quality of service will determine energy source mixes. The market will make these decisions, rather than 50 state government bureaucracies, overlaid by a federal technological bureaucracy with its whims, fancies, social engineering and pork barrel constituencies.

This transition will not be without pain. In considering state retail restructuring legislation, state legislatures[3] will be bombarded by various interest groups. Existing sinecured structures (utilities and regulators) will resist change, of course, because feudal boundaries are at stake. More substantive are the issues that arise from the built in conflicting economic interests of ratepayers, old equity holders (utility company), and new equity holders (NIPPs). These will have to be compromised-out fairly. The "securitization" mechanism for handling stranded assets offers a basis for calibrating and negotiating these interests, and proceeding forward.

Let us take, as an example, Northeast Utilities (NU), a once great utility company, now in ruins. In the past ten years, since CEO Lelan Sillin stepped down in the mid 1980s, NU has become the "Typhoid Mary" of the utility sector. Stated frankly, NU's management, post-Sillin, lacked competence and integrity. It is among the highest in the nation in electricity rates. Its reliability of service is among the worst[4]. Thus NU's customers are irked. NRC has shut down – or, rather, disallowed NU to restart-- the three Milestone nuclear units, causing the utility to have to import expensive power, and to incur expensive remediation programs. The stock has dropped from the mid-to-high 20s to as low as 7 dollars per share. Thus, the stockholders are irked. In a rather passive reactionary way, the utility is at war on multiple fronts: with NRC; with the ratepayers; with the stockholders; with the state governments of Connecticut, New Hampshire, and Massachusetts.

All of this turbulence occurs at the moment in history when electrical power sector is undergoing a fundamental paradigm shift --- toward wholesale and retail restructuring and deregulation. This fact is an opportunity, not an added burden, for the utility. The utility was essentially passive during the Connecticut State legislature formulation and debate this year on retail restructuring legislation, legislation that would have advantaged NU stockholders in a retail deregulation era. The legislation, remarkably backed both by the Governor and the Attorney General, from opposing political parties, failed.

---

[3] The Federal Government has delegated to the States the difficult details of utility retail restructuring and deregulation.

[4] In Simsbury, Connecticut, the location of our principal office, we backup our computer work every thirty minutes for fear of an outage.

The NU response to all of its self-inflicted traumas has been to blame others, and remain passive. It passed blame to NRC, a fact that did not endear them to that body, and boxed in politically what had become a relatively streamlined agency. It passed blame to its own people; the just-retired (1997) CEO hired and fired or reassigned several new nuclear executives in an effort to put distance between himself and the issues. Finally and belatedly, the Board of Directors, which itself is suspect for having these crises go unattended for years, mustered up its meager courage and handed the CEO this year a handsome retirement plan in order to remove him. NU continues to announce optimistic Milestone re-start schedules, and then fail on them. The current schedule (the latest of a failed series of promises) for Milestone Unit 3 re-start is first quarter 1998. A more realistic schedule is third quarter 1998. The utility's credibility is nil with the financial community, with its customers, with its stockholders, with the political bodies in three states, and with the NRC.

Now I ask you --- where is the ghost of Adam Smith, where is the market taking this company? I believe NU is headed either to a bankruptcy filing or to the forced divestment of the Milestone units to an entity that can manage them competently, or both.

That is where the pain comes in. The interests of the stockholders – ill served by management—will be at odds with the ratepayers, and the solution will come out somewhere in the middle. Some portion of stranded asset investment may be recovered through legislation, if the utility has the wit and leadership to work constructively and effectively with the Governor, the Attorney General, and the Legislature. The Milestone units would be spun off to a NIPP, which would acquire the asset at some value less than the book undeprecitiated value less the stranded asset recovery. The NIPP must have enough strength to operate the units effectively and profitably, and sell power throughout the grid.

That, in my opinion, is where Mother Nature will take NU. The business of generating and transmission and distribution is simply too much for NU's management talents and the sense of accountability its Board of Directors.

In the discourse above, I have used NU as an example. But this pattern will be repeated again, and again, and again throughout the country--- incompetent and inert and arrogant management, a passive Board, stockholder and ratepayer losses, leading to painful restructuring in the context of retail power restructuring.

But let me ask you this: is this not a golden opportunity for a resurgence of nuclear power in the U.S. --- to compete fairly in the market place, unfettered by a dysfunctional federal nanny who wants to call the shots and wants to get in everyone's face about everything, unfettered by a plethora of state rate regulating authorities populated largely by failed real estate and divorce and traffic claim lawyers. Is this not a vision for U.S. nuclear power, to set this entire dysfunction aside, and have the same retail access as MCI, TCI, and the other players in the telecommunication industry are developing? Make nuclear power a business, and unharness it from all of the federal, state governmental clutter and utility management incompetence. Make nuclear power compete, arms-length, on an even playing field. It can compete, but it requires government policy stability and resilient operating institutions focussed sharply on power generation.

The best prospect for a nuclear renaissance in the U.S. is through utility sector restructuring initiatives now getting underway. Utility sector restructuring itself is a sub-set of larger mega-trends: the inexorable march of history away from excessive government intrusion and dysfunctional bureaucratic baggage in both private and public sectors, toward free competitive markets, and democracy. These mega-trends get interrupted from time to time by entrenched sinecured interests. But like the Berlin Wall, these too will come down.

NUCLEAR FUSION, WORLD ENVIRONMENT, AND THE ENERGY PROBLEM: THE PATH OF MINIMUM RISK

E. Panarella

Advanced Laser and Fusion Technology, Inc.
189 Deveault St., #7
Hull, P. Q. J8Z 1S7 Canada, and
Department of Electrical Engineering
The University of Tennessee
Knoxville, TN 37996, U.S.A.

1. INTRODUCTION

The generation of energy by fusing together the isotopes of hydrogen, namely deuterium and tritium, is an objective pursued for the past 50 years by virtually all nations. In principle, it is the same process occurring in the sun and the stars, which has naturally powered the universe for billions of years. The need for pursuing this objective on Earth is compelling. As Fig. 1 shows, if we assume that the world population stabilizes at 10 billion, consuming energy at 2/3 of the U.S. 1985 rate, the energy available from fossil, hydro, and non-breeder fission will suffice approximately until the year 2030. After that, the shortfall will be increasing, and must be made up by other sources. The only alternative is fusion energy.

2. WHY DEVELOP FUSION ENERGY?[1]

Here are a few of the reasons for the need to develop fusion energy:

- Fusion energy is a virtually inexhaustible source of energy.
- Fusion will produce the most energy per amount of fuel input of any known option, with only helium gas as a releasable byproduct.
- A pound of fusion fuel contains the energy equivalent of 12 million pounds of coal, 25 thousand barrels of oil, 142 million cubic feet of natural gas, or 4.5 pounds of fission fuel.
- This translates into a land use advantage with respect to solar, wind power, geothermal, hydropower, or biomass electricity production methods.

- Fusion will not release greenhouse gases associated with global warming or chemicals associated with acid rain, nor does it contribute to other forms of atmospheric pollution.
- Fusion does produce radioactivity in the structural materials in the power plant, but its environmental impact can be reduced through choice of low activation materials, such as vanadium and silicon carbide composites that results in minimum environmental impact from fusion energy.

3. THE ECONOMIC CHALLENGE OF FUSION[2]

Achievement of fusion energy is not an objective per se. It must be placed in the context of the market requirement. The economic challenge of fusion lies in the following considerations:

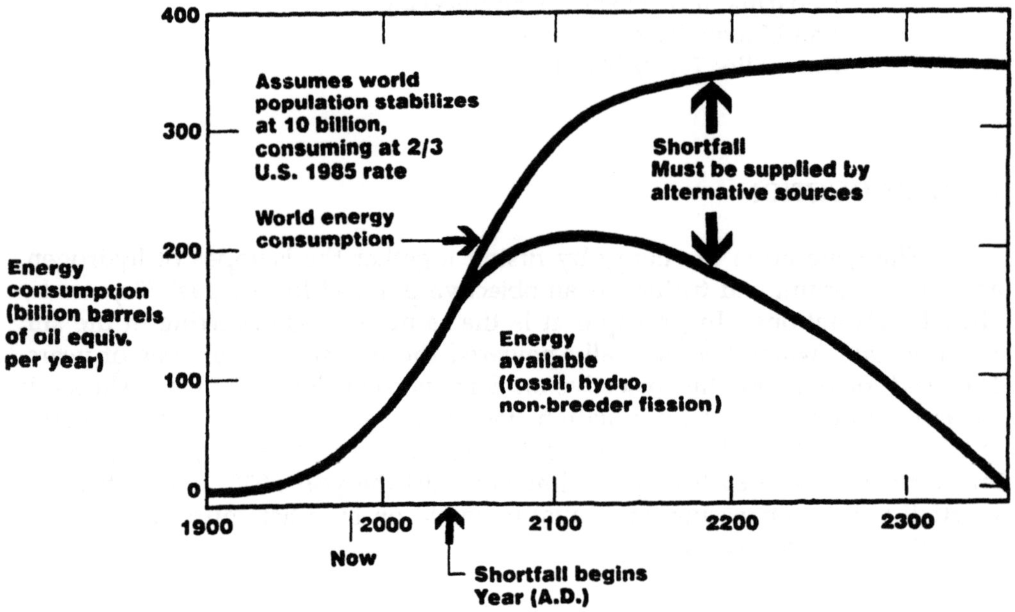

Figure 1. Energy consumption, energy available, and shortfall time.

- Fusion will be the world's ultimate energy source, if it can be harnessed economically.
- Pulsed compact tokamaks and glass lasers will continue to be powerful tools for advancing fusion science. But "economics" is replacing "scientific feasibility" as the definition of success.
- Even with its environmental superiority, fusion must respond to its customers if it is to compete in the marketplace of the 21st century. For magnetic fusion energy, the tokamak has been proposed as a start. Is it the best that fusion can offer?
- What direction should fusion take in order to increase the chance of success and lead to a substantially better reactor product?

4. THE PRESENT APPROACH TO FUSION[3]

Figs. 2, 3, and 4 show drawings of the ITER (International Termonuclear Experimental Reactor) device, a machine that has been proposed as a reactor of the magnetic confinement type to demonstrate the scientific and technological feasibility of fusion power. It is a machine of about 30 m diameter, and 36 m height. To accommodate it, a facility of ~ 30,000 m^2 for a total cost of ~ U.S. $\$10^{10}$ will be required. Its construction should begin around the year 2000, with completion by about 2010. An intense debate has been going on within the fusion community about the desirability of building such a machine. Specifically, the magazine *Physics Today* reported in its June 1996 issue an opinion piece titled "Build the International Thermonuclear Experimental Reactor?", in which Andrew M. Sessler and Thomas H. Stix on one side, and Marshall N. Rosenbluth on the other side, debated the issue. This opinion piece best represents the two sides of the issue. The arguments were as follows.

Arguments Against ITER (by Sessler and Stix)

- The ITER machine would be genuinely huge.
- The toroidal field coils would be 17 m high.
- The plasma volume would be 20-40 times larger than that of today's largest tokamaks.
- Construction time has been estimated at 10-12 years.
- Construction costs, in 1996 dollars, has been estimated at $10 billion.
- Although the magnetically confined plasmas most closely approaching fusion conditions have been produced in tokamaks, an eventual fusion reactor will look very different.
- Proceeding with ITER may not only preempt funding for alternative concepts, but also could freeze reactor design and engineering at a premature stage.
- Finally, the construction of ITER would be a poor choice for the following reasons:

    - The machine's untested attributes and unresolved technical issues.
    - Questions of science policy, including the narrowed focus on conventional tokamak.
    - Diversion and dedication of great human, scientific and fiscal resources to this single project.
    - The 10 or more years to "first plasma".
    - The time periods typical for fusion innovation and for political change of will.
    - The negative impact on future research that would rebound from a mechanical or physics failure in this single device.

In short, Sessler and Stix conclude, the step is too large and the overall concept, for all its attractiveness, is both premature and over ambitious with respect to current knowledge.

Arguments in Favour of ITER (by Rosenbluth)

- The nonlinear physics and novel engineering issues of fusion are so complex that only a real experiment at the approximate parameters required for ignition will ever resolve them quantitatively.

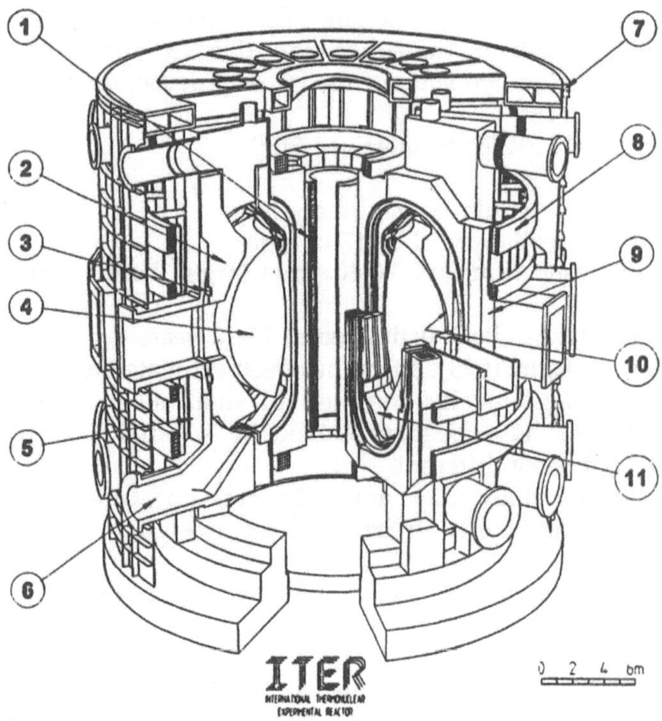

1- CENTRAL SOLENOID
2- SHIELD/BLANKET
3- ACTIVE COIL
4- PLASMA
5- VACUUM VESSEL-SHIELD
6- PLASMA EXHAUST
7- CRYOSTAT
8- POLOIDAL FIELD COILS
9- TOROIDAL FIELD COILS
10- FIRST WALL
11- DIVERTOR PLATES

Figure 2. The ITER device.

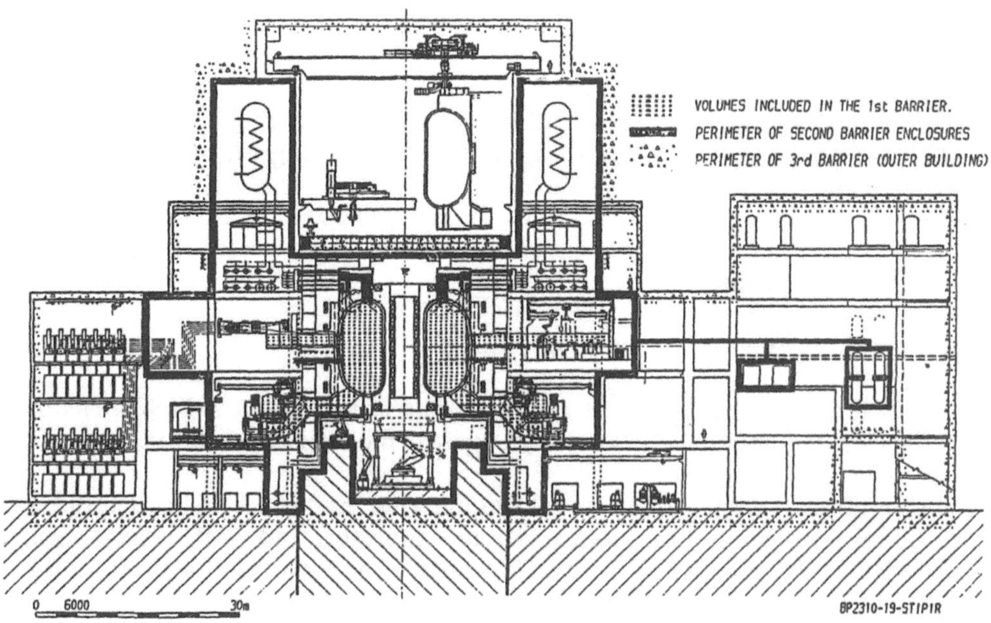

Figure 3. The ITER facility (vertical cross section).

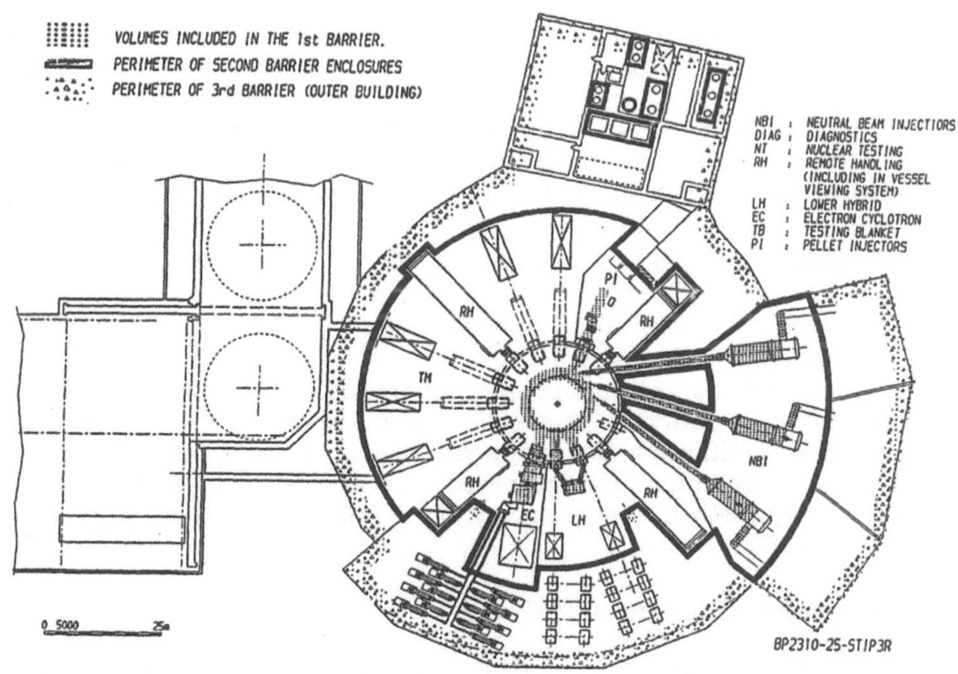

Figure 4. The ITER facility (horizontal cross section).

- Fusion research requires a test bed such as ITER that could be used to design a desirable reactor by interpolation rather than continual extrapolation from undersized experiments.
- If Japan assumes the major financial role in constructing ITER, with U.S. support at about its present level of 20% ($55 million) of the annual U.S. fusion budget, this is a wonderful bargain, because, for 5% of the cost, the U.S. could participate in designing and experimenting on fusion's flagship experiment.
- The U.S. fusion community should not turn its back on an internationally agreed upon experiment designed as the first exploration of the burning plasma experiment.

Reply to the Arguments in Favour of ITER

Since the present author is not in favour of building the ITER machine, the following arguments will be offered against those of Rosembluth.

- The nonlinear physics and novel engineering issues of fusion should be resolved with theoretical/computational analyses, and with modest scale experiments of the size of a desirable reactor, and not with a machine like ITER, that can never become a desirable reactor.
- The so-called "undersized" experiments on tokamaks have involved machines, such as JT-60, TFTR, and JET, that can be already considered gigantic machines of cost on the order of a few billion dollars each.
- Irrespective of the amount of money involved, a general financial rule that applies to any expenditure is that what is purchased must be good. If for 5% of the cost of the ITER machine one gets something that is not good, it is better to save that 5% of the money, irrespective of the fact that the deal seems to be a bargain. The ITER machine, because it will end up, in the most optimistic scenario, as an undesirable reactor, will constitute a loss of 5% of money.
- The "internationally agreed upon experiment" was never "agreed upon" by the international fusion community, but by a narrow group of fusion physicists, without real support from the fusion community at large.

5. SCIENTIFIC ARGUMENTS AGAINST ITER

There are valid scientific arguments that show that the ITER machine is flawed. Instabilities[4], and turbulence[5] have been mentioned as significantly serious to negatively affect the operation of ITER. In this section, I shall give an account, from a different perspective, of the scientific reason why ITER might not reach ignition for the design parameters of this machine. Although the calculations presented here are based on a simple model of the machine, they are nonetheless sufficiently convincing to indicate that the results do not bode well for ITER.

The design parameters of ITER are:[6]

Temperature profile	20 kV at center decreasing linearly towards the periphery
Driver power	400 MW
Particle density	10^{14} cm^{-3}

Major radius	8 m
Minor radius	2.8 m to 4 m

The most elementary definition of ignition is that plasma condition for which bremsstrahlung and conduction losses cannot cool the plasma because the α particles heating is able to compensate for these losses and maintain the plasma at a constant temperature. We shall refine this definition later on.

One can verify if ignition is achieved by analyzing the temporal evolution of the plasma through the following relation:

$$aW(t) + P_{\alpha}(t) = H(t) + R(t) + C(t) \quad (1)$$

where W(t) is the power available from a driver to heat the plasma core, *a* is the efficiency of coupling the driver power to the plasma core, $P_{\alpha}(t)$ is the α particle power, H(t) is the rate of heat energy change in the plasma core, and R(t) and C(t) are the rate of energy loss by radiation and heat conduction, respectively. Other losses, such as synchrotron or inverse compton radiation production, are not included here.

Let us consider the case when one has a driver of maximum power W_{max}. As soon as it is turned on, the driver power will rise within a certain time until W_{max} is reached, after which it remains constant. At that time equilibrium is established, the plasma will have reached a certain constant equilibrium temperature, and P_{α}, R, and C will also have constant values. However, H = 0, because the plasma has now constant temperature. During this steady-state condition we have:

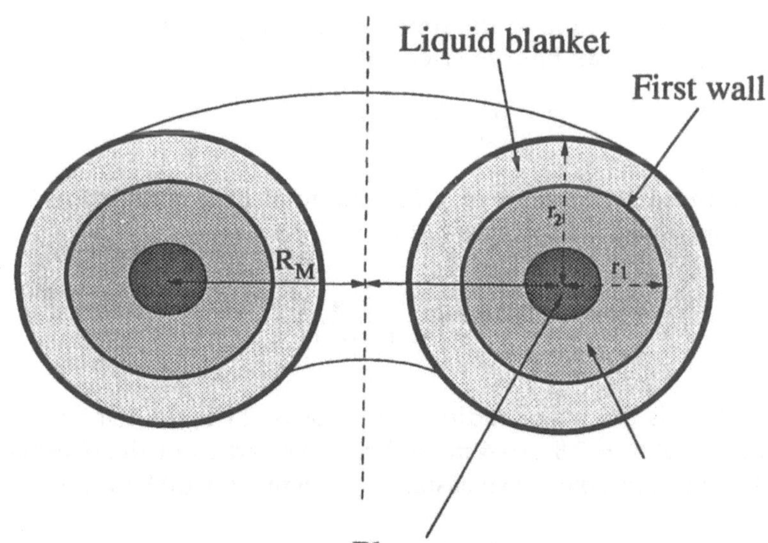

Figure 5. Simplified configuration of the ITER machine.

$$aW_{max} + P_\alpha = R + C. \tag{2}$$

Eq. (2) is satisfied for any steady-state driver input W_{max}. However, ignition is not achieved with a driver input for which $P_\alpha < R + C$. Only when

$$P_\alpha = R + C \tag{3}$$

ignition is achieved, and one can turn the driver power off at that time. It is not advisable to keep the driver power W_{max} on after ignition, because the reactor might be damaged from the combined action of W_{max} and P_α.

The objective of this analysis is not to derive from (1) the conditions to be satisfied in order to reach ignition. Rather, it is to verify if, for the ITER design parameters provided above, ignition can be achieved. In other words, since it has been claimed that, for the design parameters listed above, ITER should reach ignition,[6] we first want to derive from (2) some additional operating parameters of ITER, and then we want to proceed to verify if ignition can be achieved for those parameters.

Determination of the Temperature of the First Wall

The following analysis aims at deriving the temperature of the first wall (Fig. 5). We assume that the first wall is surrounded by a liquid blanket which absorbs whatever heat leaks out from the plasma, as well as the neutron and bremsstrahlung powers (the bremsstrahlung absorption in the first wall converts into heating of the latter with heat transfer to the liquid blanket). For simplicity the liquid blanket is considered to be made up of water at 300° C, whose thermal conductivity is:[7]

$$K_{LB} = 6.27 \times 10^4 \frac{W}{cm \cdot keV} \tag{4}$$

Eq. (2), which is valid at the time immediately before the driver power is turned off at ignition, can be converted into:

$$C = P_\alpha - R + aW_{max} \tag{5}$$

In the toroidal geometry of ITER, the total heat conduction loss rate C flowing into the liquid blanket is given by:[8]

$$C = 4\pi^2 R_M \frac{K_{LB} T_1}{\ln\left(r_2/r_1\right)} \tag{6}$$

where R_M = 8 m is the major radius of the torus, T_1 is the temperature of the first wall located at r_1 = 2.8 m - 4 m, and r_2 is the radius of the external wall of the liquid blanket container. We assume r_2 = 6 m. Eq. (5) becomes:

$$4\pi^2 R_M \frac{K_{LB} T_1}{\ln(r_2/r_1)} = \int_0^{V_p} (p_\alpha - r_B) dV_p + aW_{max} \qquad (7)$$

where V_p is the plasma volume, and p_α and r_B are the α particle and bremsstrahlung power per unit volume, respectively.

Let us convert the integral in (7) over volume V_p into an integral over minor radius r_1 of the torus:

$$dV_p = 4\pi^2 R_M \, r \, dr$$

$$4\pi^2 R_M \frac{K_{LB} T_1}{\ln(r_2/r_1)} = 4\pi^2 R_M \int_0^{2.8 \text{ m to } 4 \text{ m}} (p_\alpha - r_B) r \, dr + aW_{max}$$

$$\frac{K_{LB} T_1}{\ln(r_2/r_1)} = \int_0^{2.8 \text{ m to } 4 \text{ m}} (p_\alpha - r_B) r \, dr + \frac{aW_{max}}{4\pi^2 R_M} \qquad (8)$$

Inserting in (8) the known numerical values of the various parameters, one can get the temperature T_1 (in keV) of the first wall:

$$T_1 = \frac{\int_0^{280 \text{ cm}} (p_\alpha - r_B) r \, dr + 1.27 \times 10^4 \, a}{8.23 \times 10^4} \qquad \text{for } r_1 = 2.80 \text{ m} \qquad (9)$$

$$T_1 = \frac{\int_0^{400 \text{ cm}} (p_\alpha - r_B) r \, dr + 1.27 \times 10^4 \, a}{1.55 \times 10^5} \qquad \text{for } r_1 = 4.00 \text{ m} \qquad (10)$$

The α particle power and bremsstrahlung power per unit volume for a 50% D-T mixture are given by the following expressions, respectively:[9]

$$p_\alpha = 2.06 \times 10^{-24} \frac{n^2}{4} T(r)^{-(2/3)} e^{-\frac{19.94}{T(r)^{1/3}}} \qquad (11)$$

$$r_B = 2.14 \times 10^{-30} \frac{n^2}{4} T(r)^{1/2} \qquad (12)$$

where p_α and r_B are expressed in watt/cm³, T(r) is the plasma temperature in keV, and n is the particle density in cm⁻³.

Inserting (11) and (12) in (9) and (10), with $n = 10^{14}$ cm⁻³, one gets:

$$T_1 = 3.04 \times 10^{22} \int_0^{280\text{ cm}} \left[2.06 \times 10^{-24} T(r)^{-(2/3)} e^{-\frac{19.94}{T(r)^{1/3}}} - 2.14 \times 10^{-30} T(r)^{1/2} \right] r\,dr$$

$$+ 1.54 \times 10^{-1} a$$

$$\text{for } r_1 = 2.80 \text{ m} \qquad (9')$$

$$T_1 = 1.61 \times 10^{22} \int_0^{400\text{ cm}} \left[2.06 \times 10^{-24} T(r)^{-(2/3)} e^{-\frac{19.94}{T(r)^{1/3}}} - 2.14 \times 10^{-30} T(r)^{1/2} \right] r\,dr$$

$$+ 8.19 \times 10^{-2} a$$

$$\text{for } r_1 = 4.00 \text{ m} \qquad (10')$$

The plasma temperature T(r) is a function of the minor radius r of the torus, starting at 20 keV in the central core and decreasing linearly towards the periphery, where it reaches T_1 at the first wall.[6] Hence:

$$T(r) = 20 - \frac{20 - T_1}{280} r \qquad \text{for } r_1 = 2.80 \text{ m} \qquad (13)$$

$$T(r) = 20 - \frac{20 - T_1}{400} r \qquad \text{for } r_1 = 4.00 \text{ m} \qquad (14)$$

and (9′), (10′) transform, respectively, into:

$$T_1 - 1.54 \times 10^{-1} a$$

$$= 3.04 \times 10^{22} \int_0^{280\text{ cm}} \left[2.06 \times 10^{-24} \left(20 - \frac{20 - T_1}{280} r\right)^{-(2/3)} e^{-\frac{19.94}{\left(20 - \frac{20 - T_1}{280} r\right)^{1/3}}} - 2.14 \times 10^{-30} \left(20 - \frac{20 - T_1}{280} r\right)^{1/2} \right] r\,dr$$

$$\text{for } r_1 = 2.80 \text{ m} \qquad (9'')$$

$$T_1 - 8.19 \times 10^{-2}\, a$$

$$= 1.61 \times 10^{22} \int_0^{400\ \mathrm{cm}} \left[2.06 \times 10^{-24} \left(20 - \frac{20 - T_1}{400} r\right)^{-(2/3)} e^{-\frac{19.94}{\left(20 - \frac{20 - T_1}{400} r\right)^{1/3}}} - 2.14 \times 10^{-30} \left(20 - \frac{20 - T_1}{400} r\right)^{1/2} \right] r\, dr$$

for $r_1 = 4.00$ m (10″)

Eqs. (9″) and (10″) can be solved numerically for T_1 as a function of a. Table 1 reports the results.

Table 1. Temperature of the first wall as a function of the energy transfer efficiency a from driver to plasma, for two minor radia, i.e., $r_1 = 2.8$ m and $r_1 = 4.0$ m.

T_1 (keV)	a	
	$r_1 = 2.80$ m	$r_1 = 4.00$ m
0.181	**0.996**	-
0.15	0.796	-
0.111	0.543	**0.997**
0.10	0.472	0.860
0.09	0.407	0.739
0.08	0.343	0.617
0.07	0.278	0.496
0.06	0.213	0.374
0.05	0.148	0.252
0.04	0.084	0.131
0.03	0.019	0.008

In the above Table we have highlighted the first wall temperature when $a \approx 1$. It is 0.181 keV for $r_1 = 2.80$ m, and 0.111 keV for $r_1 = 4.00$ m.

Verification of Ignition

Although the most general definition of ignition is the one provided by Eq. (3), a more appropriate one for ITER stems from the consideration that this machine is designed to be a proof-of-principle reactor. Therefore, the reaction product, namely the neutron output power, should be taken into account in the energy balance equation. This is because this power will help in achieving the goal of the reactor, which is production of electricity.

Since the neutrons escape from the plasma and are absorbed by the liquid blanket, together with the heat and bremsstrahlung radiation losses, the power available from the liquid blanket W_{LB} for conversion into electricity is:

$$W_{LB} = P_n + R + C \tag{15}$$

where P_n is the neutron power.

A fraction b of this power is converted into electricity, where b is typically 30%. Hence:

$$W_{electr.} = b\,(P_n + R + C) \tag{16}$$

ITER is supposed to be able to reach ignition. If this is so, at ignition $W_{max} = 0$, and Eq. (2) becomes:

$$W_{max} = \frac{1}{a}\left(R + C - P_\alpha\right) = 0. \tag{2'}$$

The electricity production, if any, from ITER is obtained by subtracting (2′) from (16):

$$W_{electr} = b\,(P_n + R + C) - \frac{1}{a}\left(R + C - P_\alpha\right) \tag{17}$$

This general expression is justified because, if indeed ignition is achieved, the second term on the right-hand side of (17) is equal to zero, and (17) reduces to (16).

Equation (17) transforms into:

$$\frac{W_{electr}}{b} = P_n + \frac{1}{ab} P_\alpha + \left(1 - \frac{1}{ab}\right)(R + C) \geq 0 \tag{18}$$

where $P_n = p_n V_p$, and p_n (neutron power per unit volume) is given by:[9]

$$p_n = 8.31 \times 10^{-24} \frac{n^2}{4} T(r)^{-(2/3)} e^{-\frac{19.94}{T(r)^{1/3}}} \tag{19}$$

By inserting (6), (11), (12), and (19) in (18), one has:

$$4\pi^2 R_M \int_0^{2.8\,\text{m to }4\,\text{m}} \left[\left(8.31 \times 10^{-24} + \frac{2.06 \times 10^{-24}}{ab}\right)\left(\frac{n^2}{4} T(r)^{-(2/3)} e^{-\frac{19.94}{T(r)^{1/3}}}\right) + \left(1 - \frac{1}{ab}\right)\left(2.14 \times 10^{-30} \frac{n^2}{4} T(r)^{1/2}\right) \right] r\,dr$$

$$+ 4\pi^2 R_M \left(1 - \frac{1}{ab}\right) \frac{K_{LB} T_1}{\ln\left(r_2/r_1\right)} \geq 0 \tag{20}$$

Table 2. Values of the left-hand side of inequality (20), as a function of *a* and *b*, for two minor radia, i.e., $r_1 = 2.8$ m and $r_1 = 4.0$ m.

T_1 (keV)	*a*		*b*	Left-hand side of (20)	
	$r_1 = 2.80$ m	$r_1 = 4.00$ m		$r_1 = 2.80$ m	$r_1 = 4.00$ m
0.181	**0.996**	-	0.3	-4.91×10^8	-
0.15	0.796	-	0.3	-5.73×10^8	-
0.111	0.543	**0.997**	0.3	-6.77×10^8	-3.73×10^7
0.10	0.472	0.860	0.3	-7.13×10^8	-1.02×10^8
0.09	0.407	0.739	0.3	-7.34×10^8	-1.41×10^8
0.08	0.343	0.617	0.3	-7.56×10^8	-1.93×10^8
0.07	0.278	0.496	0.3	-7.87×10^8	-2.45×10^8
0.06	0.213	0.374	0.3	-8.20×10^8	-2.91×10^8
0.05	0.148	0.252	0.3	-8.46×10^8	-3.42×10^8
0.04	0.084	0.131	0.3	-8.59×10^8	-3.78×10^8
0.03	0.019	0.008	0.3	-8.87×10^8	-3.00×10^8

Inserting in (20) the values of *a* and T_1 reported in Table 1, and of T(r) as given by (13) or (14) for $n = 10^{14}$ cm^{-3}, and carrying out the operations on the left-hand side of the above inequality, one finds that the inequality is not satisfied, and one always gets negative numbers, rather than positive numbers, or at most zero. In (20) we have assumed $b = 30\%$. Table 2 reports the results.

Since the above table shows that ignition is not achieved, this means that Eq. (2′) is not zero but greater than zero. Despite this, one can still verify if ITER can reach power breakeven, i.e., $W_{electr} = W_{max}$, provided, of course, that the driver power W_{max} is kept on. Subtracting (2′) from (16), one gets:

$$W_{electr} - W_{max} = b\,(P_n + R + C) - \frac{1}{a}\left(R + C - P_\alpha\right) \qquad (17')$$

which becomes, for power breakeven:

$$\frac{W_{electr} - W_{max}}{b} = P_n + \frac{1}{ab} P_\alpha + \left(1 - \frac{1}{ab}\right)(R + C) = 0 \qquad (21)$$

Eq. (20) is the same as (18), and the negative numbers of Table 2 show that ITER might not even reach power breakeven for the design parameters provided before.

THE PATH OF MINIMUM RISK FOR FUSION

The above analysis shows that the ITER path for fusion is very risky, as well as expensive, and should be reconsidered. A healthy fusion program should rely on a good mix of conventional, and alternate concepts, in order to diversify the risks involved in the approach to the solution of the fusion problem. The tokamak program should not be abandoned, provided it is centered around reasonable size machines, such as the Ignitor machine[10], which promises to reach ignition at a fraction of the cost of ITER. At the same time, the alternate concepts should be vigorously pursued. To quote just a few of these concepts deserving of support, on should mention the dense Z-pinch, magnetized target fusion, muon catalyzed fusion, plasma focus, reversed field pinch, spherical pinch, spheromak, staged Z-pinch, etc. Moreover, one should leave the door open to exploratory concepts, which might very well prove to be decisive in the fusion problem. All of these concepts have recently been presented in a conference dedicated to an open view in the fusion arena.[11]

CONCLUSIONS

The present study has provided on overview of the main direction of the fusion program in the world, which is presently centered around the ITER (International Thermonuclear Experimental Reactor) machine, and found that serious doubts persist about its validity in general, and its scientific merit in particular. An analysis carried out on a simple model of the machine has shown that ITER might not be able to reach, for its present design parameters, either ignition or power breakeven. The large cost and the long time frame of the program, on the other hand, leave no room for any doubt about its success. Consequently, as long as these doubts persist, and cannot easily be dissipated, the ITER program should be put on hold, and fusion alternative concepts should be pursued.

REFERENCES

1. From Dr. Martha Krebs testimony - *Hearing before the Subcommittee on Energy of the Committee on Science, Space and Technology* - U.S. House of Representatives, August 23, 1994 (No. 14) p. 14-15.

2. L. John Perkins, James H. Hammer, and R. Paul Drake, in *Current Trends in International Fusion Research* (ed. Emilio Panarella), Plenum Press, New York (1997), p. 365.

3. Andrew M. Sessler, Thomas H. Stix, and Marshall N. Rosenbluth, Build the International Thermonuclear Experimental Reactor?, *Physics Today,* June 1996, pp. 22-25.

4. B. Coppi, F. Bombarda, P. Detragiache, S. Migliuolo, and L.E. Sugiyama, Evolution of global modes and magnetic reconnection in fusion burning plasmas, in *Proceedings of the 17th IAEA Fusion Energy Conference,* Montreal, Canada, 7-11 October 1996.

5. James Glanz, Turbulence may sink titanic reactor, *Science* 274:1600 (1996).

6. L. John Perkins, ITER Team (private communication).

7. Lindon C. Thomas, *Fundamentals of Heat Transfer*, Prentice-Hall, Englewood Cliffs, N.J., 1980, p. 653.

8. L.R. Ingersoll, O.J. Zobel, and A.C. Ingersoll, *Heat Conduction*, McGraw-Hill, New York, 1948, p. 38.

9. S. Glasstone, and R.H. Lovberg, *Controlled Thermonuclear Reactions*, D. Van Nostrand, New York, 1960, pp. 20-32.

10. Francesco Pegoraro, Ignition Physics and the Ignitor Project, in: *Current Trends in International Fusion Research*, Emilio Panarella, ed., Plenum Press, New York (1997), p. 125.

11. Emilio Panarella, *Current Trends in International Fusion Research*, Plenum Press, New York (1997).

CHARACTERIZATION AND INVENTORIES OF NUCLEAR MATERIALS AND WASTES FOR POSSIBLE FUTURE ENERGY SCENARIOS

Edward D. Arthur

Nuclear Materials and Stockpile Management
Los Alamos National Laboratory
Los Alamos, New Mexico 87545

ABSTRACT

Awareness of the total materials inventory and materials balance associated with differing methods for energy generation is part of present-day concerns associated with disparate areas that include atmospheric emissions, resource utilization, health effects, and both current and long-term hazards and risks. Nuclear energy, for a number of decades, has been the recipient of significant scrutiny concerning the materials and wastes it generates, particularly in the context of long-term solutions to such issues. This paper examines the nuclear materials and waste generation for nuclear energy scenarios spanning the coming century. The paper also briefly addresses wastes (in the form of emissions) from other energy sources and examines requirements associated with back-end energy system materials management. Possible future requirements pertaining to CO_2 management are found to place conditions upon waste management generally similar to those for nuclear waste. One example of material flows for the case of coal generation of electricity coupled with carbon sequestration is also given.

INTRODUCTION

Nuclear energy is believed by many to be an important resource for meeting future energy needs and as an important technology for greenhouse (and other emissions) mitigation. At the same time, "back-end" fuel cycle issues associated with nuclear materials and nuclear waste production are a major area of controversy across disparate topics ranging from solutions to nuclear waste disposal to near- and long-term proliferation concerns. This paper examines the back-end materials implications from several scenarios developed for future nuclear energy use worldwide. Because the area of managing

Environment and Nuclear Energy
Edited by Kursunoglu *et al.*, Plenum Press, New York, 1998

materials, emissions, and wastes from all forms of energy production has been receiving increased emphasis, the paper also briefly addresses the area of CO_2 production and management. In doing so, parallels to challenges associated with nuclear waste management are drawn along with a discussion of material flows and balances associated with electricity production using coal-fired technology and subsequent carbon sequestration.

APPROACH

The present paper uses results from two recent analyses of possible future nuclear energy scenarios. The first is part of a Los Alamos National Laboratory Nuclear Vision project analysis of global nuclear energy and materials scenarios. Reference 1 provides details on the models used, the analysis, and results obtained. Some pertinent information is provided here. The analysis of *Krakowski, et al.*[1] used the ERB (Edmonds, Reilly, Barnes) global energy-economics-environment model[2] to which a detailed nuclear fuel cycle description was added to replace the simplified model present in the ERB. The ERB model computes energy demand for fifteen-year time steps out to the year 2095 and then calculates market shares of a variety of energy sources (including nuclear energy). This is done for each of thirteen world regions plus a global result. The addition of a complete nuclear fuel cycle model allows calculation of a variety of nuclear materials inventories including spent fuel, plutonium, uranium resource requirements, etc. for a choice of several current and future nuclear fuel cycles. These include light-water reactors (LWRs) operating in a once-through fashion, reprocessing of LWR spent fuel to recover plutonium for use in mixed oxide fuels (MOX) in LWRs, plus advanced systems based upon reactor- or accelerator-based fast neutron spectrum burner systems. For the discussion here, most results will be provided for LWRs operating in a once-through mode since this approximates spent fuel management scenarios based on projections of current and near-term reprocessing and MOX fuel fabrication capacities.

The second analysis used results reported at the International Atomic Energy Agency Symposium on Nuclear Fuel Cycle and Reactor Strategy: Adjusting to New Realities, "Global Energy Outlook", Key Issue Paper 1[3]. This analysis in turn used results from other sources -- energy and nuclear energy projections from the International Energy Agency's *World Energy Outlook*[4] for the period up to 2010 and International Institute for Applied Systems Analysis (IIASA) and World Energy Council projections[5]. The IAEA study combined multiregional results from these studies into global, developed (OECD), developing, and reforming economy (former Soviet Union countries) categories. Finally, the IAEA study used a simplified add-on nuclear materials model to determine nuclear materials and waste projections as a function of time. This model incorporated evaluations of reactor-type mixtures, fuel burnup assumptions, and capacity factor projections to create such data as well as to assess impacts of differing fuel cycle and management strategies on them.

As illustrated in Reference 1 (through a number of comparisons associated with areas such as total energy and nuclear energy demand scenarios), there appears to be reasonable consistency in projections for global and the other combined categories listed above among the references cited above. Figure 1 summarizes scenarios for nuclear energy demand over the 21st Century using results from the Los Alamos analysis of Reference 1. As indicated, they span business-as-usual, low, and high scenario variants. Again, all of these are in rough agreement with projections (out to 2050) from the IAEA study and its sources. The magnitude of upward or downward changes in nuclear energy demand occurs

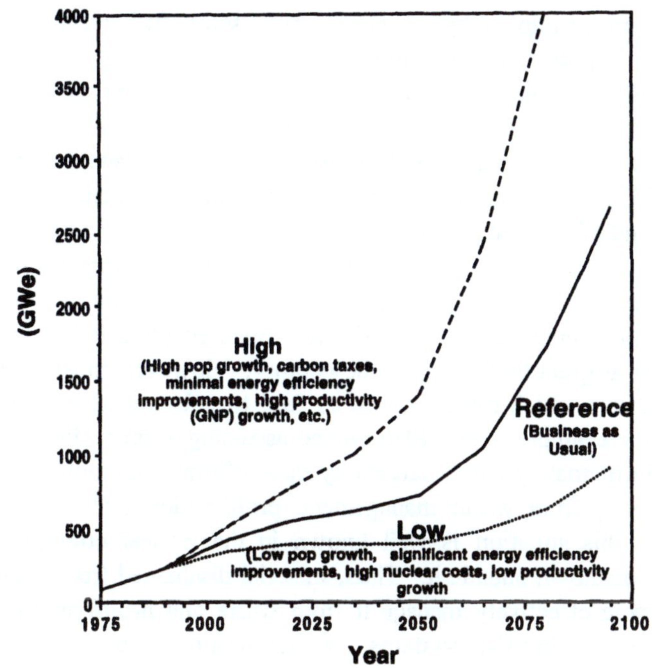

Figure 1. Global Nuclear Energy Scenarios (Reference 1)

from reasonable changes in assumptions concerning population growth, implementation of tax structures for mitigation of carbon emission from fossil fuels, energy efficiency improvements, GNP growth, nuclear costs, etc. (The sensitivity of nuclear scenario projections to such assumptions are discussed more fully in Reference 1.) One final note concerns projections for nuclear energy demand beyond 2050. All three scenarios show an upturn in demand. This mirrors projections of the ERB (and other models) for total energy demand brought about by increasing world populations in developing regions and the increase in developing countries' GNP, which in turn creates demand for increased energy, even when efficiency improvements are factored in.

The results from this analysis also indicate a shift in where nuclear energy could be expected to be implemented over the course of the 21st century. Throughout the time period to 2095, nuclear energy demand in developed (OECD) and reforming economy countries is expected to stay roughly comparable to today's levels. (In this analysis, OECD country demand totals stay roughly constant because declines expected in the United States (and perhaps Europe) are offset by increases in Asian OECD members (Japan, Korea). However, developing or industrializing countries, particularly China, account for the majority of nuclear energy capacity beyond 2050.

NUCLEAR MATERIAL PROJECTIONS

Figure 2 provides scenarios for the global inventory of plutonium in spent fuel assuming no recycle and reactor fuel burnup values of 33,000 MWd/tonne. On a global scale, such scenarios indicated plutonium growth from the current total of approximately 1000 tonnes to values between 15,000 and 40,000 tonnes by the end of the century. These figures can be changed by implementing technologies and operational practices providing increased plutonium burnup or by adopting, universally, fuel cycle strategies aimed at plutonium reuse (and potentially breeding).

For growth of plutonium and spent reactor fuel inventories, there is less of a shift in the location or country type versus time as compared to the nuclear demand discussed previously. This occurs because the backlog or "legacy" of spent fuel and plutonium occurring in OECD countries continues to dominate time-dependent inventory projections until almost the end of the 21st century.

The inventory increase of spent fuel expected globally from such projections (for the "reference" case of Figures 1 and 2) is illustrated in Figure 3. Global spent fuel inventories are large (being on the order of a few hundred thousand tonnes) and grow to amounts ten times as great by the end of the century. However, most of this inventory is comprised of uranium which can be reused as feed to subsequent re-enrichment in LWR fresh fuel. For the example shown (determined assuming a spent fuel burnup of 33,000 MWd/tonne), approximately 2 to 3 percent by mass of the total amount of spent fuel is of interest (or concern) from waste management, proliferation risk, and energy recovery points of view. This situation is well known in the nuclear community but is often unappreciated or unknown outside it. Thus, material discharged from nuclear reactors can be recycled to more effectively manage it from waste and proliferation points of view. Even without use of breeder systems, a significant amount of additional energy (approaching fifty percent) could be recovered if economic or other conditions warranted it.

The IAEA study[3] also examined the fuel cycle strategy impact of recycling on the global accumulations of discharged reactor fuel. Their scenarios included fifty and one-hundred percent recycle of discharged LWR fuel via use of plutonium-containing MOX fuel in one-third LWR cores. Their projections indicated that discharged fuel inventories could be reduced by twenty-five to fifty percent over amounts expected in 2050 without recycle. Annual natural uranium requirements could also be reduced by up to twenty percent by 2050 assuming full recycle of discharged reactor fuel.

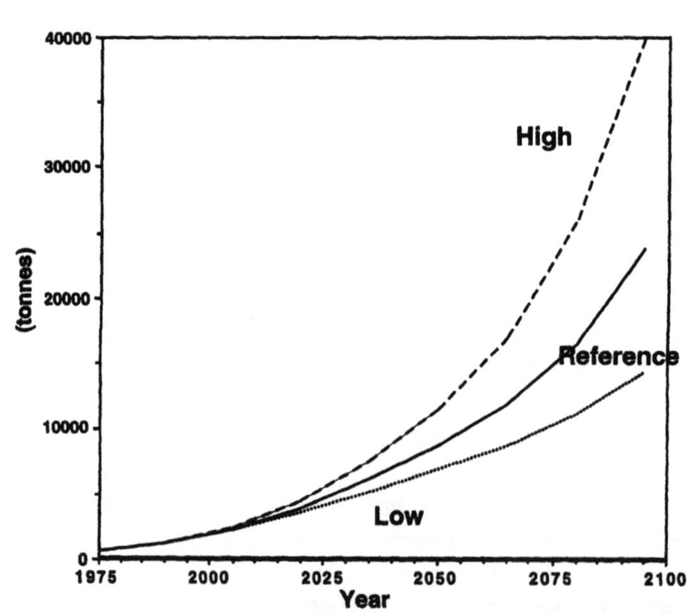

Figure 2. Global Plutonium Inventories in Spent Fuel (No Recycle Case of Reference 1)

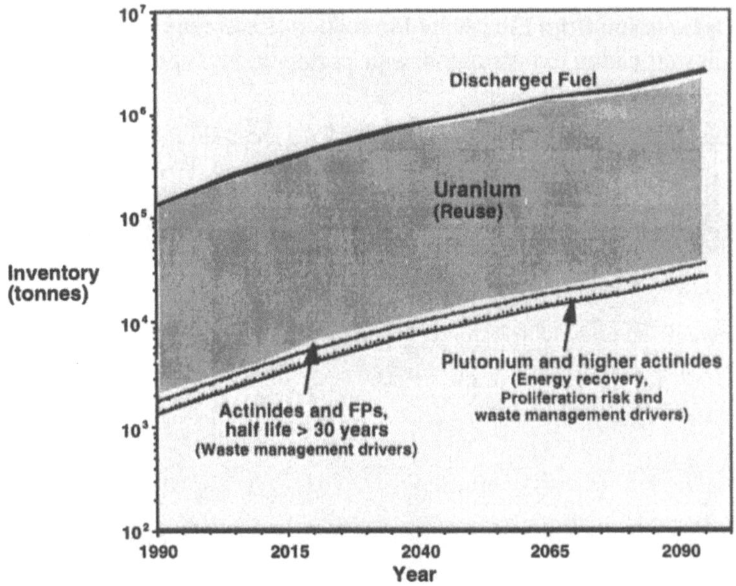

Figure 3. Discharged Fuel and Nuclear Materials Projections

COMPARISON OF MATERIAL FLOWS AND WASTES DISCHARGED FROM OTHER ENERGY SOURCES

The issue of global climate change impacts resulting from energy production is receiving increased attention both nationally and internationally. Results[6] from the ERB model indicate that the annual rate of CO_2 emission could increase by close to a factor of four before the end of the next century. Cumulative CO_2 emissions could double over the same timeframe. Starting around 2025, rapidly developing nations such as China could dominate CO_2 production, rising to a rate close to five times greater than the United States or other currently industrialized countries by 2100. Life cycle CO_2 emissions from energy source technology are therefore receiving significant attention in addition to approaches based upon CO_2 emission penalties (carbon taxes on fossil fuels), energy use and conversion efficiency improvements, and development of advanced, renewable energy sources. A number of studies have catalogued life cycle CO_2 emission; Table 1 summarizes results from Reference 7.

If future conditions require that CO_2 be sequestered, then three general approaches could be pursued -- dilution into oceans, storage using physical barriers, and combining CO_2 with calcium or magnesium oxide to create mineral carbonates. CO_2 storage involving physical barriers requires that such barriers be maintained indefinitely. Otherwise, the danger exists from an accidental release similar to the 1986 phenomena at Lake Nyos, Cameroon[8] where a bubble on the order of 0.1 km^3 (corresponding to the weekly output of a gigawatt electric power plant) was released.

A number of proposals are being developed directed towards carbon dioxide disposal in carbonate minerals. One of them[9] is briefly discussed here, primarily from the point of view of material flows associated with coal-fired electricity production rather than a critique of this particular approach for CO_2 sequestration. To fuel a one gigawatt (electric) power plant requires on the order of 190,000 tons per day of earth resulting in 9000 tons per day of coal to feed the plant. Approximately 24,000 tons of CO_2 are

Table 1. CO_2 Emission from Electricity Production (Reference 7) (Includes contributions from fabrication, construction, transportation, and operation).

Electric Power Plant Type	CO_2 Emission (g-carbon/kWhr)
Coal	237 (combined cycle) - 274
Liquid Natural Gas	108 (fuel cell) - 180
Oil	204
Nuclear	7 - 9
Geothermal	11
Wind	9
Solar	40

produced daily along with other emissions (depending upon the efficiency of scrubbers and other techniques used to reduce NO_x and SO_2 emissions). In the proposal of Reference 10, approximately 50,000 tons per day of rocks rich in magnesium and calcium oxide would be required for the sequestering process in addition to significant daily amounts of other materials used in the process.

This discussion indicates that present technologies for energy and electricity production often have significant materials movement, generation, and waste creation. For nuclear energy, material flows and discharged reactor fuel arisings can still be viewed as significant, but their amounts are much lower than for other energy sources. Storage of such materials also occurs in a finite number of sites (initially on-site at reactor sites followed by interim storage at a much smaller number of more centralized discharged fuel storage facilities, with plans for eventual disposal in several geologic sites in various world regions).

CONCLUSIONS

Projections from several published scenarios of nuclear energy demand trends occurring in the 21st century have been used to determine associated inventories of discharged reactor fuel and inventories of key materials occurring in these discharged reactor stocks. Such projections indicate that "demographics" (location of nuclear power plants, type of countries where nuclear development could be expected to occur, etc.) could be significantly different over the next fifty years of nuclear power than over the past several decades. Inventories of discharged reactor fuels and plutonium in them are projected to increase by at least an order of magnitude over the next fifty to one-hundred years. This increase may serve, to some observer, as furthering the intractable nature often associated with long-term management and disposal of nuclear materials and wastes. However, current and future technological approaches based upon recycle of discharged fuel could reduce expected inventory growth by at least a factor of two (for more conventional methods of recycle) and potentially much more if advanced technologies are developed and used. Decisions on whether to do so will, however, result from a mixture of technical, political, and institutional factors that include issues associated with the overall economics of nuclear power, its acceptability, and perspectives on future proliferation dangers (real or perceived) that may be associated with civilian nuclear energy.

At the same time, the life-cycle material and waste generation from other energy sources will be subject to increasing scrutiny as more attention is paid to the coupling of energy demand, its generation, and regional and/or global environmental impacts. In particular, expected wide-spread use of coal for electricity production in developing countries will involve extremely large amounts of material flows and waste production. Additionally, efforts to mitigate emissions from the use of such fuels, particularly those aimed at dealing with CO_2 control, could add significantly to such material flows.

REFERENCES

1. R.A. Krakowski, J.W. Davidson, C.G. Bathke, E.D. Arthur, and R.L. Wagner, Jr., "Nuclear Energy and Materials in the 21st Century", International Symposium on Nuclear Fuel Cycle and Reactor Strategy: Adjusting to New Realities, IAEA, Vienna (June 3-6, 1997).

2. J. Edmonds and J.M. Reilly, *Global Energy: Assessing the Future*, Oxford University Press, New York, NY (1985).

3. H.F. Wagner, P. Jelinek-Fink, L.L. Bennett, and E. Bertel, "Global Energy Outlook", Key Issue Paper Number 1, International Symposium on Nuclear Fuel Cycle and Reactor Strategy: Adjusting to New Realities, IAEA, Vienna (June 3-6, 1997).

4. International Energy Agency, *1996 World Energy Outlook*, Organization for Economic Cooperation and Development, Paris (1996).

5. N. Nakicenovic (Study Director), "Global Energy Perspective to 2050 and Beyond", World Energy Council and International Institute for Applied Systems Analysis report (1995).

6. R.A. Krakowski, "The Role of Nuclear Energy in Mitigating Greenhouse Warming", Los Alamos National Laboratory Report, LA-UR-97-4891 (December 1997).

7 S. Yasukawa, Y. Tadokoro, and T. Kajiyama, "Life Cycle CO_2 Emission from Nuclear Power and Fuel Cycle System", Proceedings Expert Workshop on Life-Cycle Analysis of Energy Systems, Methods, and Experience, International Energy Agency, Paris (May 21-22, 1992), p151.

8. S.J. Freeth, "Lake Nyos: Can Another Disaster Be Avoided?", Geochemical Journal **28**, 163 (1994).

9. K.S. Lachner, D.P. Butt, C.H. Wendt, "Progress on Binding CO_2 in Mineral Substrates", Los Alamos National Laboratory Report, LA-UR-96-2818 (August, 1996).

SECTION III Nuclear Energy: Economic, Political, and Environmental Considerations

REDUCING CO_2 EMISSIONS: TOWARD A SUSTAINABLE ENERGY MIX

William F. Martin
Former U.S. Deputy Secretary of Energy
Former Executive Secretary, National Security Council
Board Member, World Resources Institute
October 28, 1997

INTRODUCTION: CONTEXT FOR THE CO_2 EMISSIONS CHALLENGE

Scientific debate is highlighting the link between CO_2 and climate change.

IPCC Conclusions. In a report released in late 1995, the Intergovernmental Panel on Climate Change (IPCC), which represents the majority of the world's scientists, asserted that "the balance of evidence suggests a discernible human influence on global climate." More specifically, the IPCC predicted that "if no changes are made in emissions of greenhouse gases, scientific analysis projects an increase in global mean surface air temperature relative to 1990 of 2°C by 2100 . . . this rate of warming would be the greatest experienced in the last 10,000 years." Though the debate continues, even these tentative statements signal the possibility of an energy and environmental future that is less than secure.

Facing the Rio Commitment. The magnitude of the risk revealed by the IPCC's first report led more than 150 countries to sign the Framework Convention on Climate Change (Rio Convention) in Rio de Janeiro in May 1992. The Convention became legally binding in March 1994, after ratification by the required 50 signatories. The majority of industrialized countries adopted specific targets to limit carbon dioxide or total greenhouse gas emissions; about half adopted the target to stabilize and/or reduce anthropogenic emissions of carbon dioxide by 2000 or 2005 to levels prevailing in an earlier year. Despite these announced targets, with the exception of the United Kingdom, no industrialized country will meet its greenhouse gas emission targets by 2000.

> "The balance of evidence suggests a discernible human influence on global climate change . . . If no changes are made in emissions of greenhouse gases, scientific analysis projects an increase in global mean surface air temperature relative to 1990 of 2°C by 2100 . . . this rate of warming would be the greatest experienced in the last 10,000 years."
>
> *IPCC Second Assessment Synthesis of Scientific-Technical Information Relevant to Interpreting Article 2 of the UN Framework Convention on Climate Change, December 1995*

Environment and Nuclear Energy
Edited by Kursunoglu *et al.*, Plenum Press, New York, 1998

The Background Established at Berlin. In the spring of 1995, the Conference of Parties under the Framework Convention held its first meeting in Berlin. Asserting the inadequacy of the Rio Convention, delegates set new emission reduction targets in the "Berlin Mandate" in 2005, 2010, and 2020. At a later meeting in 1996, the United States agreed to accept legally binding targets for reducing emissions. These legally binding targets would apply to nations that are members of the Organization of Economic Cooperation and Development (OECD), but not developing nations.

The Challenge at Kyoto. The Kyoto Conference is the fourth and final international summit to be held under the auspices of the Rio Summit. Delegates from over 150 nations will gather in Kyoto, Japan, to decide upon the next set of targets. Leading up to the conference, intense debate over carbon dioxide buildup and its impact on global climate change has resulted in mixed conclusions. Developed and developing nations alike are watching these issues evolve, looking for cues to shape their actions as the 21st century approaches. Amid the uncertainty characterizing the discussion even today, one certainty is the recent and growing concern arising over the effects of human activity on the environment.

Seeking the Right Energy Mix. While recognizing that uncertainty exists, the nations of the world should contémplate the possibility that carbon dioxide may be threatening the global environment. In considering this scenario, a conscientious global community will want to determine which fuel mixes can lower carbon dioxide emissions while providing sustainable energy security.

THE GEMS MODEL PROVIDES A MEANS OF ASSESSING THE IMPACT OF OUR ENERGY MIX ON CO_2 EMISSIONS.

Assessing the Energy Mix. Our global energy model provides a means to assess the mix of energy sources by region through the year 2050. The Global Energy Supply and Demand Model (GEMS) was first constructed at MIT by Carroll Wilson and the Workshop on Alternative Energy Strategies, where I served as a program officer 20 years ago. It was further developed for the 1996 Trilateral Commission Study on Energy Security, projecting out to the year 2010. Using the GEMS model, we can examine the aggregate CO_2 contribution from different energy consumption scenarios. We can use the results generated by the model to analyze the consequences for energy policy in the developed and developing countries.

An Assumption-Driven Model. The GEMS model is driven by a set of assumptions concerning the rates of economic and population growth, fuel preferences, and energy intensity. Economic growth on a per capita basis is adjusted for the increase in population, yielding an average annual increase of 0.7 percent or smaller. By combining the rate of economic growth with the energy intensity of an economy, the model predicts future energy demand. The GEMS model uses conservative estimates of economic growth, indicating that energy demand could be even higher than the model assumes.

Inner Workings of the GEMS Model. The model matches supply with demand in the industrial, transportation, and residential and commercial sectors. Each sector has a "default" primary energy source which accounts for any demand not met by other sources. For example, transportation relies upon oil, and electricity generation relies upon coal.

An Environmental Tool. Developed to predict potential energy security crises, GEMS can also project the future environmental impact of our energy dependence. To determine the impact on CO_2 emissions of changes in the energy mix over the next half-century, we applied to the GEMS model four case scenarios involving shifts in energy supply and demand:

1. The Business as Usual Scenario: CO_2 emissions continue to rise as the world maintains heavy dependence on oil and coal. Natural gas and renewables play a small, increasing role in global energy supply, having little impact on global

warming. Nuclear energy use declines. This perpetuation of an unstable fuel mix threatens future energy security.

2. The Conventional Clean Energy Scenario: Increased reliance on natural gas and renewables achieves some reduction of CO_2 emissions, but not enough to abate the threat to the global environment. Energy security continues to remain an issue, despite reduced dependence on Persian Gulf oil.
3. The Nuclear Inclusion Scenario: Resolution of nuclear waste, safety and proliferation concerns enables generators to retire inefficient coal-fired plants and contributes to significant additional CO_2 reductions. Aggregate fossil fuel consumption remains high, however, and carbon dioxide levels still border on unsustainable.
4. The Sustainable Future Scenario: This scenario incorporates the supply-side measures of increased renewables, natural gas, and nuclear energy with demand-side conservation and energy efficiency measures. This combination of long term supply-side balance and demand-side management would optimally diversify the fuel mix, control greenhouse gas emissions, and reduce global dependence upon the volatile Persian Gulf.

PART I: FOUR GEMS MODEL CASE SCENARIOS

In the business as usual scenario, continued dependence on oil and coal threatens global security and environment.

Oil Profile. In the business as usual scenario, we assume that oil consumption will rise from 3060 to 7000 million metric tonnes per year. It will remain the exclusive fuel in the transportation sector as vehicle-miles-traveled (VMT) increases exponentially due to the release of repressed Asian demand. This increase will be only slightly offset by a continued withdrawal of oil from the electricity sector in favor of coal and natural gas. The

Table 1. 1990 Primary Energy Supply and Consumption Matrix (in million metric tonnes of oil equivalent)

	Coal	Oil	Natural Gas	Nuclear	Renewables	Electricity	**Total**
Indigenous Consumption	2322	3236	1702	515	248		**8024**
Exports/Imports (Net)	-19	175	-22	515	0	-0.60	**-217**
Total Primary Supply	**2302**	**3060**	**1680**	**-515**	**248**	**-0.60**	**7807**
Energy Sector (Electricity)	-1292	-515	-675	0	-248	833.52	**2413**
Total Final Consumption	**1011**	**2359**	**943**	**0**	**0**	**804.36**	**4797**
Transportation	18	1272	9	0	0	17.63	**1316**
Industrial	769	515	552	0	0	423.90	**2261**
Commercial/Residential	223	249	382	0	0	362.82	**1218**

Source: GEMS Global Energy Supply and Demand Model

higher demand will be met in part by production in the North Sea, the Caspian region, Russia, and the extraction of heavy oil from Venezuela and Canada. The majority of this rise in oil demand, however, can only be satisfied by the Persian Gulf oil producers. This increased dependence on such a politically unstable region will further expose the global economy to the potential of another energy security disaster.

Coal Profile. The business as usual scenario assumes that the percentage of coal in the global energy mix will rise from 2300 to 6750 million metric tonnes of oil equivalent per year. Developing countries such as China and India will continue to rely on coal for electrical generation. This increase in usage will heavily impact regional and global climate change and worsen growing environmental concerns. Already, acid rain and carbon dioxide emissions problems in regions such as East Asia have surfaced, and increased coal use will only deepen these effects.

Natural Gas Profile. Under the business as usual scenario, we assume that natural gas use will rise moderately from 1680 to 3300 million metric tonnes of oil equivalent per year. However, such an increase will require extensive infrastructure investment, which will prove difficult to accomplish.

Nuclear Profile. We assume a downturn in nuclear energy use, as there are no new plant orders in the U.S. and Europe, and many of the existing plants are scheduled for retirement. The void created by the closed nuclear plants will most likely be filled by coal-fired generators which release, on average, 5,000 lbs of CO_2 per ton of coal.

Renewables Profile. According to the business as usual scenario assumptions, there will be a continued increase in the use of renewables. However, even with this rise, renewables will not achieve serious market penetration, due to the currently small percentage fulfilled by the various renewable energy sources in the overall energy mix.

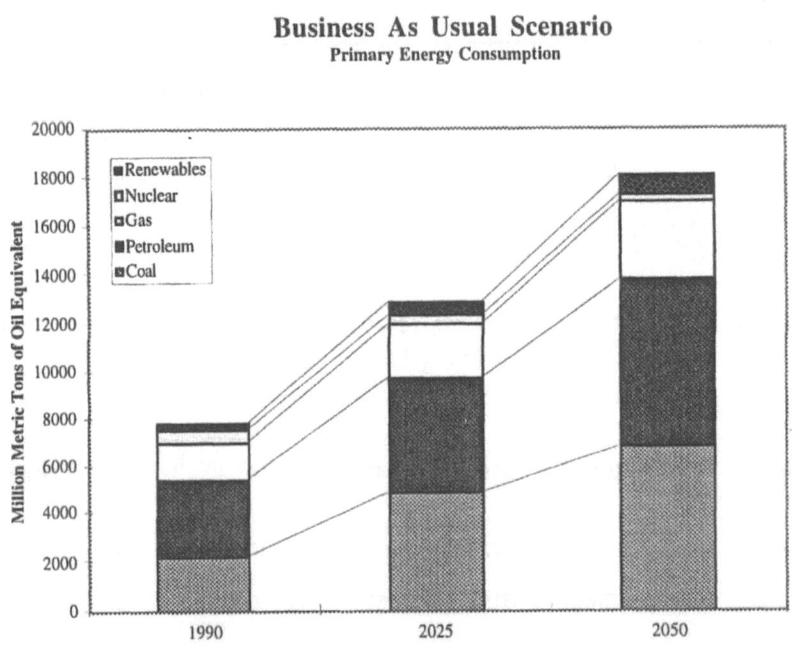

Source: GEMS Global Energy Supply and Demand Model

Figure 1. Business as Usual Scenario

The Bottom Line. Due to economics, availability, and easy transport, oil and coal will remain the dominant energy sources over the next several decades. Demand for oil would require that oil production in the Persian Gulf region increase by two-and-a-half to three times. The same factors which have made oil and coal the easy choice for industrial and commercial use until the present time will continue to provide disincentives to change for the good of the environment and a safe, sustainable energy future. The bottom line is that the business as usual scenario, including a decline in nuclear energy use, will induce a rise in carbon dioxide emissions from 26 to 62 billion tons per year.

In the conventional clean energy scenario, natural gas and renewables play a greater role in the energy mix, taking one step toward CO_2 emissions reduction.

Natural Gas Profile. In this scenario, natural gas consumption increases considerably, from 1700 to 5500 million metric tonnes of oil equivalent, due to expansion in transportation and electricity generation. New aeroturbine technology generates lower cost electricity, and increased investment in pipelines and LNG (liquid natural gas) tankers allows the product to reach the market more efficiently. The infrastructure required for considerable expansion of natural gas use would involve significant economic costs, as transportation will be complicated and expensive.

Renewables Profile. This scenario assumes that renewables become a significant contributor to the energy mix, experiencing rapid growth of ten percent until 2010, and slowing to four percent as they become more mature. Renewables total 3500 million metric tonnes of oil equivalent in 2050. These early high growth rates arise from the full realization of all hydroelectric potential and the currently small aggregate consumption of renewables. In 1990, hydroelectric accounted for 95 percent of the world consumption of renewables. By 2025, much of the hydroelectric potential will be realized, so growth will take place in solar, wind, and biomass technologies. Based on the scenario generated by the GEMS model, these sources need to increase by more than three times by 2050 if we

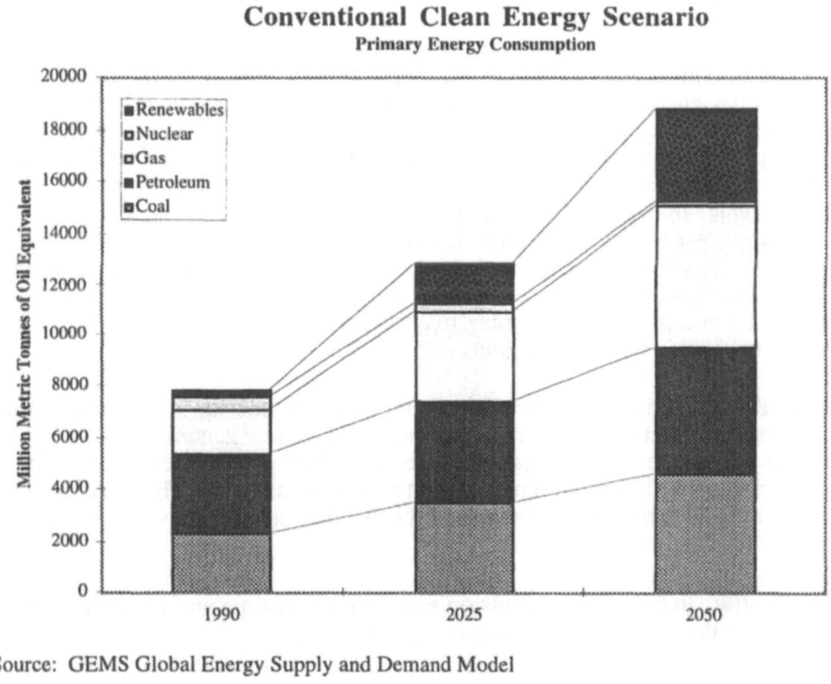

Source: GEMS Global Energy Supply and Demand Model

Figure 2. Conventional Clean Energy Scenario

hope to achieve sustainability. To accomplish this, OECD governments should adopt policies that promote investment in R&D to advance these renewable energy sources for integration with many conventional grid-based systems.

Oil Profile. The conventional clean energy scenario assumes that oil will continue to be a dominant energy source, with demand increasing to 5000 million tonnes per year. The primary reductions will be made in the transportation sector, as electric and CNG (compressed natural gas) vehicles establish a presence in the market. The increased use of alternative fuels would lessen dependence on Persian Gulf oil.

Coal Profile. Coal consumption is assumed to increase at a slower rate in this scenario, as natural gas and renewable energy sources replace some coal generating capacity. Coal remains the fuel of choice for developing countries and large centralized utilities in OECD countries, and demand for coal still rises from 2322 to 4560 million metric tonnes of oil equivalent.

Nuclear Profile. According to the assumptions in the conventional clean energy scenario, nuclear power still faces waste disposal, reactor safety, and public relations challenges. These concerns continue to impact public acceptance, inhibiting nuclear energy's ability to play an increased role in the energy mix.

The Bottom Line. In the conventional clean energy scenario, the bottom line is that CO_2 emissions would be considerably lower than in the business as usual case, due to dramatic increases in dependence on natural gas and renewables. CO_2 emissions in 2050 would reach the level of 51 billion tons. However, coal consumption continues to rise as deregulation in the United States and economic development in the industrializing nations encourage electricity producers to rely on large, centralized coal fired plants. Thus, even with increases in natural gas and renewables, continued dependence on coal and oil induces an unsustainable rise in CO_2 emissions.

In the nuclear inclusion scenario, nuclear energy overcomes public acceptance and technical challenges, contributing to CO_2 emissions reduction by replacing some coal capacity.

Nuclear Profile. This scenario assumes that opposition to nuclear power diminishes due to safer reactor design, resolution of the nuclear waste problem, and an effective public outreach campaign. CO_2 emissions would decline as nuclear plants replace some coal generating capacity. Nuclear capacity would recover and increase to 1000 GW by 2050, representing 17.5 percent of global electricity generation and a tripling of nuclear power use from current levels. In industrializing nations heavily reliant on coal, such as China, the substitution of nuclear energy could provide definitive relief from acid rain and CO_2 emissions. To facilitate safe, secure nuclear energy development, we would encourage specific nations with mature nuclear technology to actively participate in China's development of a nuclear energy program, to ensure resolution of environmental and proliferation concerns in the Asian region.

Coal Profile. Based on the model assumptions, the nuclear inclusion scenario would significantly affect coal consumption. Increased nuclear generating capacity would present a direct substitute for the large centralized coal-fired plants. Demand for coal in 2050 is limited to only 3125 million metric tonnes of oil equivalent. These factors impacting coal use account for the majority of the decrease in CO_2 emissions in this scenario.

Oil Profile. In the nuclear inclusion scenario, we assume no further decrease to oil consumption over the "conventional clean energy" scenario.

Natural Gas Profile. The scenario assumes that reliance on natural gas will remain the same in the electricity sector, but increase in the transportation sector.

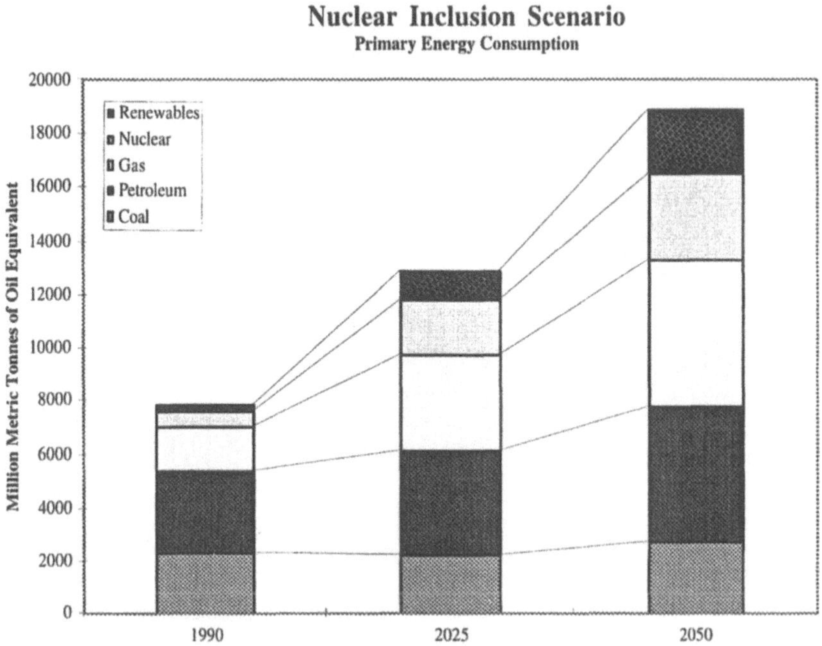

Source: GEMS Global Energy Supply and Demand Model

Figure 3. Nuclear Inclusion Scenario

Investment in infrastructure, new generating facilities, and automotive R&D will allow it to capture 29 percent of the aggregate fuel mix.

Renewables Profile. Renewable energy sources will receive the same attention as in the conventional clean energy scenario. Government-sponsored R&D in the areas of solar, wind and biomass technologies will bring the price of these technologies into the range of conventional energy sources.

The Bottom Line. A devoted supply-driven effort including increased nuclear energy use will succeed somewhat in reducing CO_2 emissions, but its impact remains limited. The GEMS model shows that even if renewables and nuclear power generation expand to meet half of the additional energy demand anticipated by the year 2050, global warming would continue to rapidly increase. Since coal releases 5000 lbs of CO_2 per ton of coal combusted, by replacing a large percentage of these facilities with nuclear power, CO_2 emissions drop significantly. The bottom line is that in the nuclear inclusion scenario, the substitution of nuclear generating capacity for coal-fired plants would cap total CO_2 emissions at 44 billion tonnes per year.

The sustainable future scenario implements energy supply and demand management measures, resulting in a balanced fuel mix and the most sustainable level of CO_2 emissions possible.

Conservation. The GEMS model illustrates the importance of demand-side management programs to achieving a sustainable future. The model predicts that implementation of effective energy efficiency programs can have the same effect on CO_2 levels as eliminating nearly 50 percent of the expected increase in fossil fuel emissions. The sustainable future scenario assumes efficiency savings of 25 percent, as energy efficiency rises from current levels. Energy efficiency and conservation efforts would benefit from expanded development and use of alternatively fueled vehicles and clean electricity sources to fuel them.

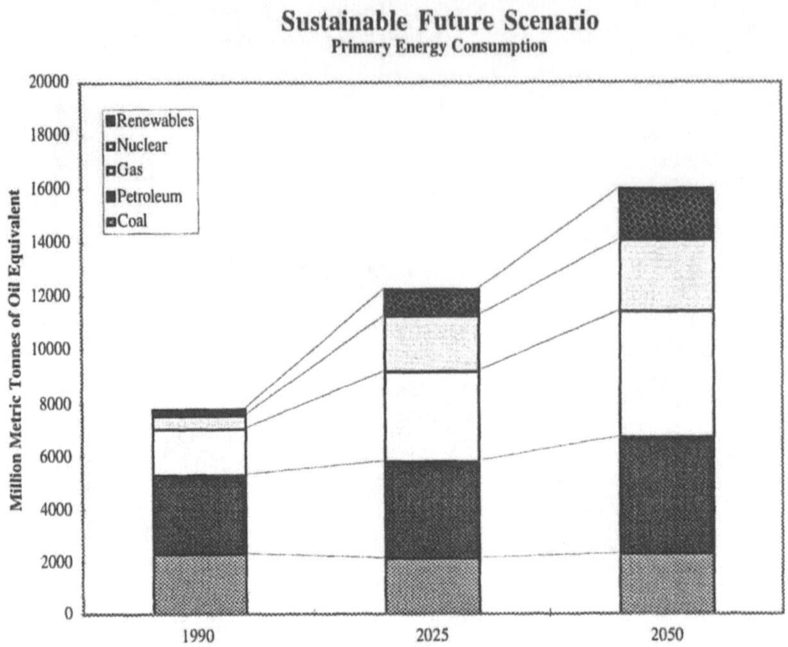

Source: GEMS Global Energy Supply and Demand Model

Figure 4. Sustainable Future Scenario

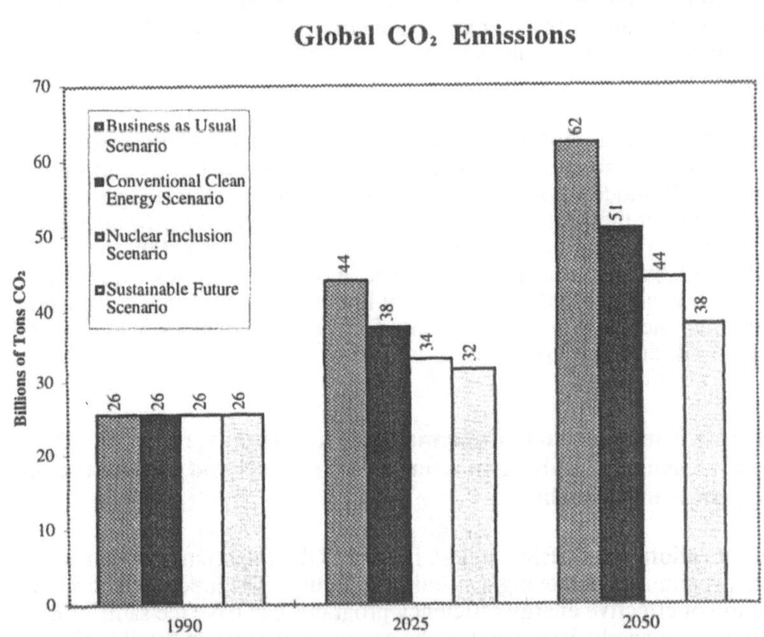

Source: GEMS Global Energy Supply and Demand Model

Figure 5. Global CO_2 Emissions

Oil Profile. In the sustainable development scenario, oil use would increase, but not much above today's levels. Oil demand growth, which rises to 4360 million metric tonnes of oil equivalent, is due to the expansion of developing country demand. Diversification of energy sources coupled with this small increase in oil demand would facilitate reduced dependence on Persian Gulf oil supplies.

Coal Profile. The sustainable development scenario would see a minimal increase in coal as a percentage of total energy demand in all sectors. For economic reasons, coal remains the primary fuel for industrializing and developing nations, but consumption is limited to 2676 million metric tonnes of oil equivalent.

Natural Gas Profile. The sustainable development scenario would require a nearly threefold increase in natural gas consumption, with expansion occurring in the electrical generation, transportation, and industrial sectors.

Nuclear Profile. As in the nuclear inclusion scenario, we assume that nuclear generating capacity reaches 1000 GW in the sustainable development scenario.

Renewables Profile. Heavy encouragement of R&D through technological advancement and financial support allows renewables to assume 28.75 percent of the electricity generation mix.

The Bottom Line. What kinds of changes in the energy mix should we envision to arrive at a sustainable development scenario over the next several decades? The key lies not in any one fuel, nor in any one demand-side measure. Achievement of a sustainable future depends on heavy market penetration by renewable energy sources, a commitment to the nuclear option, and effective efficiency programs. Most of the world's incremental energy demand will come from industrializing countries, especially in Asia and Latin America, due to a restructuring of the world economy in a post-Cold War free trade environment. If advanced nations do not take the lead in implementing cleaner, more efficient, safer energy sources and conservation measures, the unsustainable plight of the developed nations will become a global hazard. If all of these measures are implemented, the GEMS model predicts only a slight increase in carbon dioxide emission rates from today's levels, to 38 billion tons per year by the year 2050.

PART II: ASSESSING THE ENERGY MIX OPTIONS: POLICY RECOMMENDATIONS

The global community will face challenges with every energy mix option, but facing these challenges will avert the dangers of an unsustainable future

Business as Usual Equals Bad Economics. The business as usual scenario is economically flawed from a global perspective. If the shift of heavy industry from the advanced to the developing nations takes place without diversification of the fuel mix, the budding industrial economies will be indefensible when the next oil price shock occurs. Persian Gulf oil producers hold nearly the same market share as in the mid-1970s. Events in this volatile region produced the oil crises in 1973 and 1979, which then became the debt crisis of the 1980s and threatened the stability of world financial markets. If another oil shock were to occur as a result of oil demand dependent on Persian Gulf suppliers, OECD nations would likely emerge intact. In contrast, developing economies reliant on oil and coal to support their growing industries would flounder. Due to the interconnections within the global economic system, the suffering of the developing nations would transfer to the rest of the global economic community.

An Environmentally Unsustainable Trend. In the long term, the business as usual scenario is not sustainable environmentally. Continued heavy dependence on coal and oil with no real contribution from clean energy technologies will set the stage for dire

Table 2. Fuel mix recommendations for sustainable development

- *Oil*: Diversify away from the Persian Gulf suppliers.
- *Coal*: Develop clean coal technology.
- *Natural Gas*: Ensure the infrastructure for transportation, and promote international cooperation to protect the integrity of international pipelines.
- *Nuclear*: Develop advanced reactor technologies, while resolving safety, waste, and proliferation concerns.
- *Renewables*: Accelerate development of solar, wind and biomass sources.
- *Energy Efficiency*: Encourage a 25 percent increase in energy efficiency.
- *Automobiles*: Encourage alternatively fueled vehicles and energy efficiency.

environmental consequences. Even in the short term, acid rain deposition in the industrializing Asian nations is threatening to reduce agricultural output and is causing health problems for many residents. A recent study by the RAINS-ASIA program predicts that sulfur dioxide levels in the region will triple the combined output from North American and Europe by 2010. These negative health effects have become international, as pollution from coal and oil dependent developing nations spreads to neighboring countries. Looking beyond these current environmental consequences of past and present human activity, longer term global climate change concerns loom if the future does not see a concerted effort toward change.

Moving toward a National Security Threat. Despite increased production in the Caspian and North Seas, the majority of the increase in demand of this magnitude can only be met by the OPEC producers, and of that group, the only suppliers with seemingly endless oil resources are the Persian Gulf producers. Instability continues to plague the Persian Gulf region, creating an undependable oil supply. If global demand for oil continues to rise unabated, the national security costs of maintaining secure access to oil will likely become unsustainable as military leaders such as the United States become reluctant to bear the costs of sending troops to the Persian Gulf.

Steps Toward a Sustainable Future. The other three scenarios contain measures which achieve increasing levels of sustainability. Each new adaptation of the energy supply and demand matrix involves greater challenges, even as each enables a greater possibility of a sustainable energy and environmental future through the next half-century. The current trends in energy development require thoughtful consideration of all possible options, and the necessary conclusion that the optimal solution should incorporate both supply and demand side management. It is also evident that ultimate sustainability is the most difficult to attain and the most costly, given our current economic system.

In striving for the sustainable energy scenario, the global community will face challenges inherent to development of each element of the energy mix.

Oil. OECD countries should aim to completely eliminate oil from electricity generation and adopt policies encouraging market penetration of alternatively fueled vehicles. Ultimately, oil supplies should be used to meet transportation needs in developing countries. Advanced nations should assist the Former Soviet Union in achieving greater clarity in tax and investment codes. The United States can contribute through increased technology in existing fields and environmentally sensitive development of Alaskan resources. In diversifying oil sources, the global community can reduce dependence on the unstable Persian Gulf.

Coal. Coal will remain a very important energy source through the next century. As coal use also poses the greatest environmental danger, development and transfer of clean coal technologies to the industrializing nations are essential. OECD countries, recognizing that global economic success depends upon the ability of emerging economies to access low cost energy supplies, should cooperate with developing economies to secure clean coal programs worldwide. In many of these industrializing nations, coal is in abundance and already plays a dominant role in their future energy plans.

Natural Gas. Natural gas is the cleanest of all fossil fuels and has an established infrastructure in most OECD countries. As the GEMS model predicts that natural gas consumption should triple by 2050, we need to make efforts to protect the integrity of international gas transportation routes. Heavy investment will be required in developing nations, particularly Central Asia, to construct long-distance pipelines and build liquid natural gas (LNG) tankers. Eventually, there will be a need for an international panel to guarantee protection for pipelines which cross multiple borders.

Nuclear Energy. Nuclear energy needs to play a role in the energy mix through the next century to achieve sustainable development. The nuclear industry needs to guarantee reactor safety, an effective solution for waste, and non-proliferation of weapons-grade material. We should strive to alleviate proliferation and safety concerns in the Former Soviet Union. International cooperation on safe nuclear energy programs, especially in Asian nations such as China, would benefit both the global environment and global security. To meet increased nuclear scenarios, we should increase uranium consumption efficiency. Countries with mature nuclear programs and non-proliferation credentials should continue development of advanced nuclear systems and recycling technologies, including reprocessing and MOX fuel utilization. President Clinton emphasized the added merits of uranium recycling, stating that MOX fuel technology can help to eliminate surplus plutonium.

Renewables. Nations should make efforts to fully develop hydroelectric power potential, while pursuing development of other renewable energy technologies. Once hydroelectric energy's maximum potential has been realized, OECD governments should support further basic research in solar, wind, and biomass technologies in order to lower costs and increase effectiveness. Renewables should be integrated into centralized grid-based systems if they are to play an important role.

Conservation. Energy efficiency programs should be strongly encouraged in OECD and undeveloped nations. Successful conservation measures are generally difficult to implement because they require active participation by the public. They should be designed for long-term sustainability from a technical, financial, and public perspective. Countries that have managed successful programs should provide consultation to other nations. Technology transfer will prove crucial as developing nations seek guidance and assistance for safe energy development.

Achieving a sustainable future will require a broad-minded approach by all nations.

Alternatively Fueled Vehicles. Each of the four case scenarios assumes at least some movement away from oil use in the transportation sector. To maximize the environmental effects of this shift, we need to pursue development of both alternatively fueled vehicles and clean electricity sources to fuel them.

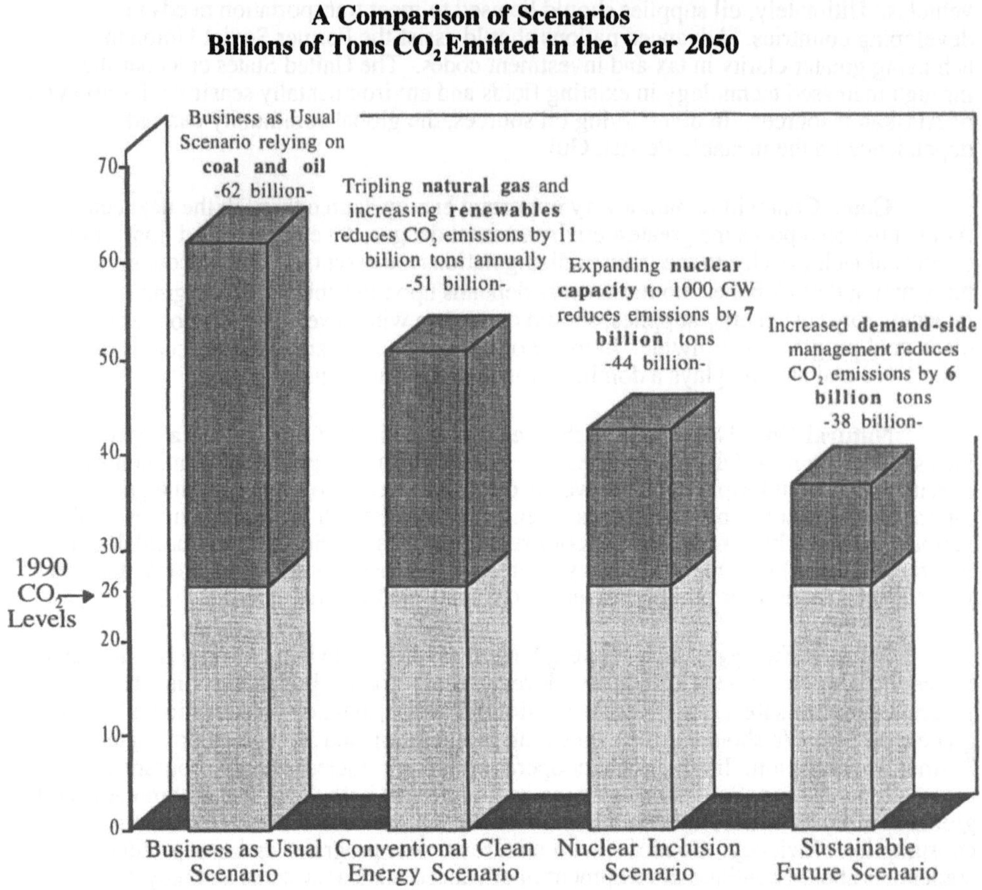

Source: GEMS Global Energy Supply and Demand Model

Figure 6.. A Comparison of Scenarios

Leadership of OECD Nations. The world needs OECD nations' leadership in seeking advanced energy technologies. Support from these advanced nations will enable developing countries to pursue development of advanced energy sources in the future. A positive relationship between developed and developing nations will foster global energy security. Steps toward such international cooperation have already begun, as the International Energy Agency (IEA) has engaged in relationships with Russia and China to exchange information and expertise on energy concerns.

Recognizing the Consequences of Inaction. The global community needs to comprehend the environmental, economic, and international security implications accompanying the business as usual scenario. Economic growth relies in part on energy development, and the pursuit of progress in these two areas often comes at the expense of the environment. A healthy outlook on the interrelationship of energy and economics will incorporate environmental concerns, as well. While the long term concern of global climate change urges the attention of policymakers, even in the short term, acid rain presents a global threat.

Evaluating the Externalities. We should translate the external costs associated with the business as usual scenario into monetary terms. These include the costs of air pollution, maintaining a military force in the Persian Gulf, and the economic depression that accompanies an oil shock. We should recognize that energy is a security issue, striving toward resolution of problems in the Middle East and the Caspian Sea. Energy can either serve as the engine of economic and security development, or it can become a political, economic, and environmental constraint. Whether energy aids or harms development depends on the energy mix we choose to implement in resolving the problems we face today.

Internalizing the External Costs. In our current economic system, the means for exacting payment for these costs are few. Conventional private and public transactions essentially discount the future relative to the present, encouraging consumption today at the expense of tomorrow. To achieve sustainability, prices should be determined by considering all accrued costs for a given time frame and working backward toward present conditions. Economic success is directly linked to the ability of countries to produce goods using an efficient, reasonably priced supply of energy. OECD nations' use of clean, advanced energy sources allows developing nations access to affordable energy sources. Competition for the same resources will only raise prices, subjecting developing economies to a debt crisis as they pay dearly for the fuels necessary for economic growth.

Appendix A: Gems Model Calculation Methodology

The GEMS Global Energy Supply and Demand Model generates energy supply and demand projections for major countries, regions and the world as a whole to the year 2050. For this presentation, GEMS has developed four scenario cases: (1) "business as usual," (2) "conventional clean technologies," (3) "nuclear inclusion," and (4) a "sustainable future" scenario. Under the "business as usual" scenario, projections are based on current practices that rely heavily on oil for transportation and coal for electrification. Each of the other scenarios consists of successively more sustainable energy mixes with reduced dependence on oil and coal, increased support for nuclear power, and decreased energy intensity.

Countries and Regions

The GEMS Model compiles energy statistics from the base year 1990 to develop energy demand projections for Canada, China, France, Germany, India, Japan, South Korea, Taiwan, the United Kingdom, and the United States. These projections are incorporated into the regional statistics compiled for Africa, the former Soviet Union, Latin America, the Middle East, Non-OECD Asia, Non-OECD Europe, OECD Europe, OECD North America, OECD Pacific and the OECD Total. The regional totals are aggregated in the world total. GEMS uses the standard energy supply/demand integration sheet used extensively by energy analysts around the world.

Assumptions

The assumptions made for each country and region include economic growth, energy growth/GDP growth ratios, oil price, indigenous production estimates, and sector demand shares.

Economic growth. The economic growth assumptions are calculated from data available from the OECD Secretariat and the World Bank. The average economic growth rate for OECD countries is projected to be approximately 1.8 percent out to 2050. In rapidly developing countries the economic growth rate is much higher, with an average of 2.5 percent out to 2050.

Energy Growth/GDP Growth Ratio. The energy growth/GDP growth ratio is one of the most important assumptions made in the model. This ratio determines the energy intensity of a country and makes the distinction between rapidly industrializing

(China and India) and industrialized (OECD) countries. Rapidly industrializing countries have a higher energy growth/GDP growth ratio because their industries (steel and heavy manufacturing) are more energy intensive. Industrialized countries have lower energy growth/GDP growth ratios because their industries (service industries) are less energy intensive. This ratio is multiplied by the economic growth rate to determine the final energy demand growth rate. The GEMS Model takes into account that the energy efficiency of technologies used by developing countries is steadily improving. While they may not always adopt the most efficient technologies, the basic standard has risen dramatically over the last 15 to 20 years, allowing developing countries to leap-frog to more efficient energy technologies. Having the benefit of more efficient technologies will result in developing countries having lower energy/GDP ratios than currently industrialized countries at a similar earlier stage of development. The "sustainable future scenario" incorporates technological improvements which result in a 25 percent savings in global energy intensity by the year 2050.

Oil Price. The price of oil is assumed to rise from $18 per barrel in 1995 to $24 per barrel in 2010 in 1995 U.S. dollars. The price is then expected to level off, increasing slightly until 2050. Oil markets are expected to remain generally steady over the period.

Indigenous Production Estimates. GEMS estimates country and regional indigenous energy production (oil, coal, and natural gas) based on figures from the OECD, the U.S. Department of Energy, and the World Bank.

Sector Demand Shares. The sector demand shares by fuel for the transportation, industrial, and commercial/residential sectors are based on the 1990 base year data. For the different scenarios, changes are made to the sector demand shares to show increased (or decreased) usage of oil, coal, natural gas, nuclear, and renewable energy sources in each sector over the sixty year period.

GEMS Model Calculation Methodology

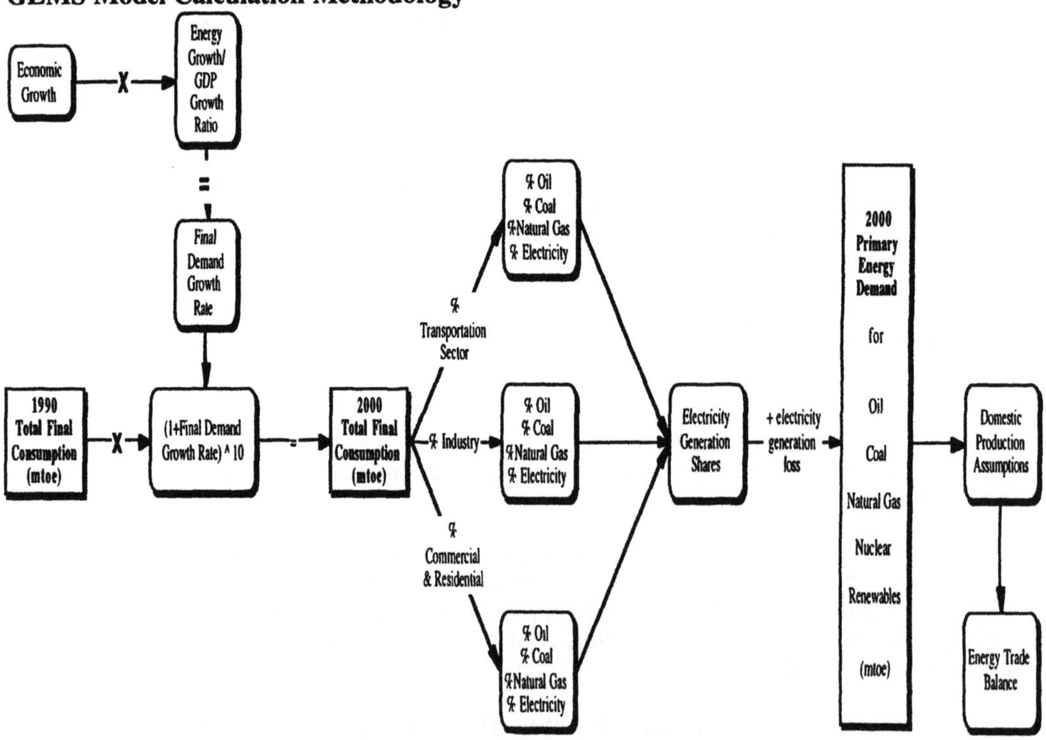

Figure 7. GEMS Model Calculation Methodology

Appendix B:
Calculation of CO_2 Emissions

Calculation of CO_2 Emissions

Data in the GEMS model is generated in a constant BTU format, consistent with that of the Secretariat of the International Energy Association. By relating all figures to the BTU equivalence of one tonne of oil, local discrepancies in calorific values between various types of coal, petroleum, and natural gas can be balanced out. The accepted standard is:

$$1 \text{ million metric tonnes of oil equivalent} = 3.968 \times 10^7 \text{ MBtu}$$

The data in the GEMS model is generated in million metric tonnes of oil equivalent. Values must be related to the short ton, equal to 2000 lbs. This can be accomplished by the following equations:

1 metric tonne = 2204.6 lbs
1 short ton = 2000 lbs

Coal. Determination of emissions from the combustion of coal is based on Table 1.1-19 of the U.S. Environmental Protection Agency Publication, AP42, Default CO_2 Emission Factors for U.S. Coals. Conversion factors are determined by the equation:

$$[44 \text{ ton } CO_2/12 \text{ ton C}] \times 0.99 \times [2000 \text{ lb } CO_2/\text{ton } CO_2] \times [72.6 \text{ lb } CO_2/\text{ton \% C}][1/100\%]$$

Where:

1. 44 is the molecular weight of CO_2
2. 12 is the molecular weight of carbon, and
3. 0.99 is the fraction of fuel oxidized during combustion

Global coal consumption assumptions:

Coking coal:	12%
Bituminous:	39%
Sub-Bitum.:	14%
Brown:	28%
Oven & Gas:	7%

[note: data source is the IEA/OECD publication: *Energy Balances of OECD Countries*]

Utilizing Emission Factor data from table 1.1-19 where available and calculating the remainder from the following equation:

$$\text{Emission Factor (EF)} = \%C \times 72.6$$

Where:

1. % C is the percentage of carbon in the various types of coals.
2. 72.6 represents an average of the values given in Perry's Chemical Engineers' Handbook of average carbon contents for each coal type (dry basis) based on extensive sampling of U.S. coals.

This generates an average aggregate emissions factor of 5160 lbs CO_2/ton coal. Carbon dioxide emissions can then be calculated by the following equation:

$$[\text{MMTOE-coal}] \times [1 \text{ tonne}/0.6514\text{toe}] \times [2204.6 \text{ lb}/1 \text{ tonne}] \times [1 \text{ ton coal}/2000 \text{ lb coal}] \times [5160 \text{ lbs } CO_2/\text{ton coal}]$$

Where:

1. MMTOE-coal is the amount of coal consumed in metric tonnes of oil equivalent

2. 0.6514 is an average conversion factor (metric tonne of oil equivalent/metric tonne) taken over specific data reported by the IEA for all major coal consuming countries
3. 2204.6 lb/tonne is the conversion factor between metric tonnes and the U.S. standard short ton

Oil: The equation used to relate carbon dioxide emissions to petroleum consumption is drawn from the EPA publication AP42, Table 1.3-11, Default CO_2 Emission Factors for Liquid Fuels. Factors are based on a 99% conversion of fuel carbon content to carbon dioxide using average fuel carbon contents listed in Perry's Chemical Engineering Handbook:

[MMTOE-oil] x [1 tonne/1.04toe] x [2204.6 lb/1 tonne] x [1/density of oil] x [Emissions Factor in lbs CO_2/gal oil]

Where:

1. 1.04 is the average of country specific conversion factors (relating Mmtoe to metric tonne) reported by the IEA
2. 7.8 lb/gal is an average density of oil

The emissions factor for oil consumption is estimated to be 24.7 lb CO_2/gal. oil

Natural Gas: The equation used to relate carbon dioxide emissions to natural gas consumption is based on the data provided in the U.S. EPA publication AP42. All factors represent uncontrolled emissions and are drawn from Table 1.4-3, Emission Factors for Carbon Dioxide and Total Organic Compounds From Natural Gas Combustion:

[Mmtoe-natural gas] x [3.968×10^7 MBtu/1×10^6 Mmtoe] x [120 lbs CO_2/1MBtu]

Where:

1. Mmtoe-natural gas is the amount of natural gas consumed globally, measured in tons of oil equivalent
2. 3.968×10^7 MBtu/1×10^6 Mmtoe is the Btu value of one ton of oil equivalent
3. 120 lbs CO_2/1 MBtu is the emissions factor for 1 Btu of natural gas

REFERENCES

Environmental Protection Agency. (1997) Publication AP42: *Compilation of Air Pollutant Emission Factors.* Washington, DC: EPA.

Intergovernmental Panel on Climate Change. (1995) *Climate Change in 1995: The Science of Climate Change.* [Draft Summary for Policy-makers of the draft contribution of Working Group I to the IPCC Second Assessment Report.] World Meteorological Organization and United Nations Environment Programme.

International Energy Agency. (1992, 1997) *Energy Balances of OECD Countries. Paris: OECD.*

International Energy Agency. (1992, 1997) *Energy Statistics and Balances of Non-OECD Countries. Paris: OECD.*

International Energy Agency. (1992, 1997) *Energy Statistics and Balances of OECD Countries. Paris: OECD.*

International Energy Agency. (1992, 1997) *World Energy Outlook.* Paris: OECD.

Martin, W.F., H. Steeg, and R. Imai. (1996) *Maintaining Energy Security in a Global Context.* New York: The Trilateral Commission.

Background on William F. Martin

William Martin served as U.S. Deputy Secretary of Energy, Executive Secretary of the National Security Council in the White House, and Special Assistant to President Ronald Reagan. In 1992, he was Executive Director of the Republican Party National Platform Committee.

During the second oil crisis, Mr. Martin was Special Assistant to the Executive Director of the International Energy Agency, Ulf Lantzke. He also held positions in the Department of State and MIT's Workshop on Alternative Energy Strategies headed by Carroll L. Wilson.

Today, he is Chairman of Washington Policy and Analysis, Inc., an international energy consulting firm. He holds degrees from the University of Pennsylvania and MIT. Mr. Martin is a member of the Board of the World Resources Institute and a member of the Council on Foreign Relations, New York.

He was recently the lead author of a Trilateral Commission study, *Maintaining Energy Security in a Global Context.* (September 1996)

INTERNATIONAL NUCLEAR UTILITY

Juan Eibenschutz

Vice President
Luz y Fuerza del Centro
México, D. F. México

As the globalization of the economy progresses, opportunities arise for new concepts in different fields of economic activity, including energy. Oil companies are one of the prime examples of multinational energy enterprises that cover a very large spectrum of organizational modalities. There are no set rules since these corporations have their own identities, the point is that among them, practically every possible way of dealing with the oil business is covered.

Depending on the specific situation of a given country, oil companies can be monopolistic or subject to full or partial competition. They may own resources or simply be allowed to exploit them. They may operate as totally owned foreign companies, or they may only operate jointly with local companies and may or may not hold the majority.

It can be said that oil companies are the best examples of global corporations, with all the flexibility to respond to different political and regulatory environments.

Opening, unbundling, deregulation, privatization, etc. of the electricity business in most countries is creating a new market for developers, equipment manufacturers, financiers, architect-engineers, power companies, etc., that are entering electricity supply within the different frameworks devised by countries in their quest to globalize and modernize.

The purpose of this paper is to propose an organizational concept that would increase the possibilities of deploying nuclear power, within the context of globalization, safety, and non-proliferation following, to a certain extent, the organizational schemes of oil companies.

The future of nuclear power

Presently the situation and the future of nuclear power can be characterized, among others, by the following facts:

- generation of about one fifth of the global electricity

Environment and Nuclear Energy
Edited by Kursunoglu *et al.*, Plenum Press, New York, 1998

- adverse public opinion
- practically no future plans except in Asia
- in spite of availability of fossil resources, nuclear energy use unavoidable, except for major scientific breakthrough
- environmental characteristics undervalued
- fears of proliferation,
- unfair economic competition, since other sources do not internalize total costs
- capital requirements significantly more intense than for other electricity producers,
- future unfavorable to nuclear because of privatization trends

Countries with significant and committed nuclear programs for the future are Japan, Korea, Taiwan, India and China. In all of them, except perhaps for China, the driving force for the deployment of nuclear power is the scarcity of other energy resources, as was the case in France and other European countries, where the deployment of nuclear power has stopped because of market saturation.

Regardless of the fact that in most western countries energy people have stopped talking about nuclear power for fear of being accused of being devils, nuclear power is playing an important role in energy supply, and should have a much more important participation in the energy markets if society is, or will be, willing to do something about greenhouse gases.

Presently the Asian countries seem to be the only ones to support nuclear power, therefore, if things continue as they are for a couple of decades, even Framatome, Siemens, AECL, ABB, General Electric, Westinghouse, etc. may loose their capacity to supply future nuclear power plants when required.

On the other hand, many smaller countries where energy resources are scarce, lack the human and material infrastructure to embark upon the nuclear power programs that would help in their economic development processes. More importantly perhaps, in the case of such countries there are concerns about nuclear safety and in some cases even about proliferation.

During the sixties some countries and some manufacturers were offering full nuclear fuel services. The main feature was that they offered to take responsibility for the irradiated fuel and to deal with radioactive residues. The former Soviet Union actually supplied its allies with this comprehensive service (mainly to assure nonproliferation).

Developments in the environment for nuclear power ended up preventing supplying countries from taking responsibility for permanently storing irradiated fuel or residues, since it became politically unacceptable to store others' wastes.

The proposed organizational concept is based on the need to keep the nuclear option alive, and the requirement to minimize both, safety hazards and proliferation risks.

The basic underlying philosophy is that the future use of nuclear power will be unavoidable, unless there is a major scientific breakthrough in energy, or unless energy is no longer required because of whatever reason.

Uranium and Thorium resources, with breeding, contain more energy than coal, gas and oil put together. It is absurd to ignore this fact, as well as the fact that nuclear power is the most environmentally friendly energy conversion process.

When energy researchers and specialists, as well as policy makers are talking seriously about fossil fired power plants equipped with carbon sequestration facilities, and when the use of biomass is seriously being considered as a long range energy solution for a technologically developed and energy consuming world, where the now developing countries have reached comparable levels of energy consumption as the ones in the richest countries, there is something very wrong.

Renouncing to the use of one of the main energy resources would be criminal for the future sustainability of mankind.

International Nuclear Utility

INU is a concept of a full energy services corporation, capable of accommodating to the particular requirements of local electric utilities or energy supply authorities.

An INU would be equipped to sell nuclear generated electricity to a grid, or to supply BLT nuclear power plants, or to take charge of the operation of existing nuclear power plants, or to produce energy vectors, such as hydrogen or methanol, at remote nuclear power facilities, in all cases taking responsibility for the complete fuel cycle, and eventual decommissioning of the facilities.

As stated previously, the concept is similar to the organizational schemes of large oil companies, except that the power of an INU could be such that it may have the capacity of politically dominating the world, since INUs would have enormous financial strength and would control huge amounts of special nuclear material.

The challenge is therefore to establish enough safeguards to prevent such a situation.

How can this be accomplished?

First, assurance of non-proliferation is more achievable since the number of reprocessing and storage facilities would be greatly reduced, and the task of the International Atomic Energy Agency (or of a similar specialized international safeguards agency) would be greatly facilitated.

Second, INUs would have partitioned ownership schemes to assure that no single group or individual could dominate the INU's decisions and operations.

Contractual arrangements to supply power or energy vectors would depend on the customer's requirements, or on the power market's characteristics. In principle, an INU could also establish subsidiaries, jointly with local utilities or energy authorities.

One possibility for the concept to materialize is that the big nuclear manufacturers establish their INUs, based on the reactor technologies they possess, and complement the spectrum of required technologies and capabilities by involving architect-engineers, fuel cycle companies, operating companies, constructors, etc.

Another modality worth of consideration is the creation of an INU that would contract with specialized companies the different supplies and services required to provide the customer's demands.

One very important feature is to improve the economics of nuclear power in order to ameliorate its competitiveness; in this respect, INUs would allow decreasing costs through mass production of standard plants, along the lines that made for the success of the French nuclear power program.

The INU would offer standard plants, with as much factory mounting (or site fabrication) as possible. It would be interesting to bring back for analysis the "barge mounted" concept, on the grounds that it could have been abandoned prematurely because of the coincidence of regulatory difficulties with the instauration of the anti-nuclear mood in several countries.

It is a generally accepted fact that petroleum technology has advanced significantly as a result of joint exploration and production joint ventures among the oil companies. The joint participation of competing companies has resulted in the greatly improved techniques, that have made possible the increases in reserves and production of oil, which were considered unattainable because of price stagnation.

This is an important lesson for the nuclear industry, one that could be applied with the INU concept.

Obviously, the ideas expressed here are very far from a framework for action, nevertheless, they warrant, in the opinion of the author, serious consideration on the part of the interested parties.

INTEGRAL ASSESSMENT OF ENERGY SYSTEMS

Stefan Hirschberg

Paul Scherrer Institute
CH-5232 Villigen PSI
Switzerland

INTRODUCTION

The electric utility sector is of central importance for economic growth and social development. While numerous societal and economic benefits arise from electricity production, it can also have negative impacts. Today, an increasing number of electric utilities are working on integrating environmental, social and economic aspects into their decision-making processes in an effort to move the electricity sector towards sustainable development. Some utilities have demonstrated that sustainable energy development is not a costly add-on to their core business; instead given cost-effective implementation it can improve their competitive position in a fast-changing industry (Boone and Howes, 1996).

In the Swiss case, due to the eventual decommissioning of the currently operating nuclear power plants and expiration of long-term nuclear electricity import contracts there will eventually open a gap between the postulated electricity demand and the available base supply (primarily hydropower). The assessment of various candidate options that could cover the prospective gap needs to be multi-disciplinary, comprehensive, reasonably detailed and, if feasible, should ultimately be carried out in an integrated manner. The commitment to sustainable development calls for the evaluation of the extent to which the different options fulfill the sustainability criteria. These criteria address use of energetic and non-energetic resources, ecological compatibility, stability and sensitivities with respect to external influences, and socio-economic aspects (Kroeger, 1998).

Efforts aiming at the comprehensive assessment of energy systems along the lines mentioned above are pursued within the Swiss research project GaBE (Hirschberg, 1995), carried out by the Paul Scherrer Institute (PSI) in collaboration with the Swiss Federal Institute of Technology Zurich (ETHZ). The multidisciplinary assessment addresses health risks, environmental impacts and economic aspects. Electricity generation, heating, transport sector and end-uses are included.

The present paper provides some highlights from the GaBE project. First, the conceptual framework for the activities is presented. Second, some selected assessment highlights are provided, illustrating the process leading to a consistent quantification of specific indicators. These examples focus on the ecological and risk-related aspects,

Environment and Nuclear Energy
Edited by Kursunoglu *et al.*, Plenum Press, New York, 1998

illustrating the corresponding advancements in the comparative assessment. Finally, the integration process is highlighted, including a pilot application of multi-criteria analysis.

OVERALL FRAMEWORK

Figure 1 shows the conceptual framework for the assessment of energy systems as used in the Swiss case. While in this paper only the electricity sector is addressed, the framework is applicable to the whole energy sector. In fact, due to many important interactions and interdependencies this broad perspective is frequently necessary. As an example, the appropriate consideration of cogeneration systems calls for the analysis of both electricity and heat markets.

The basic approach encompasses the application of Life Cycle Analysis (LCA) for each stage of the various energy chains to generate detailed material/energy flows and to establish as complete as possible set of emissions as well as other residuals.

Modeling of environmental impacts makes use of LCA and, in a parallel activity, also aims at an explicit establishment of correlations between the emission rates of air pollutants and the resulting concentrations and impacts for Switzerland. This forms a link to the assessment of health effects associated with the normal operation of energy systems. Severe accidents are being analyzed using the actual experience as well as the approach of Probabilistic Safety Assessment (PSA).

Within the energy economics a large scale model is employed to improve the understanding of the interactions between energy, economy, environment and technology. The model is used to study long-term structural changes on Swiss energy markets, to assess the importance of new energy technologies in meeting different policy goals, and to analyze the Swiss options to curb atmospheric pollutant emissions, particularly greenhouse gases.

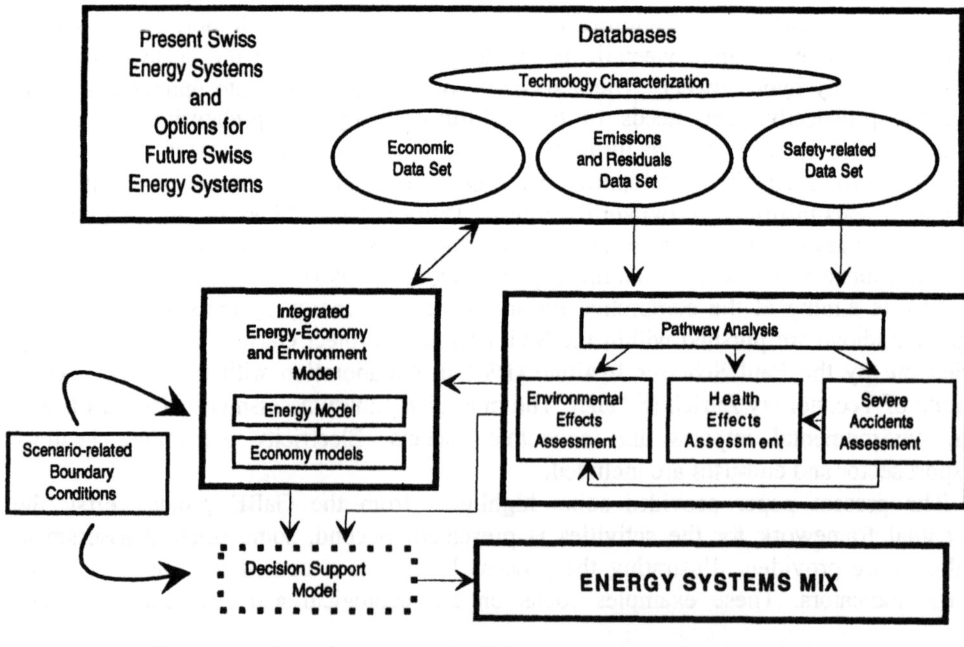

Figure 1. General framework (GaBE) for assessment of energy systems.

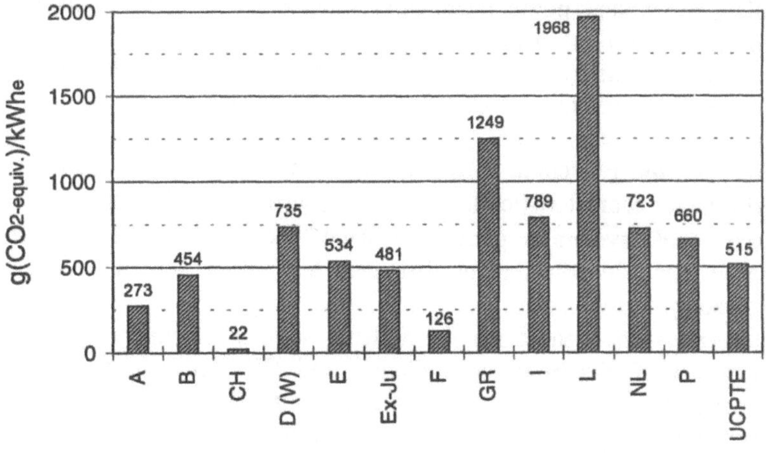

Figure 2. LCA-based GHG emissions from country-specific electricity mixes; based on (Frischknecht et al., 1996).

SELECTED ASSESSMENT HIGHLIGHTS

Environmental inventories

Figure 2 shows the normalized Greenhouse Gas (GHG) emissions from the present electricity mixes in UCPTE[1] countries. These results were obtained using Life Cycle Assessment (LCA) and are based on the new edition of ETHZ/PSI's "Environmental Inventories for Energy Systems" (Frischknecht et al., 1996).

The figure illustrates the superior GHG performance of the present, hydro- and nuclear-based Swiss electricity supply system and implies that the compatibility of major structural changes in the current configuration with the environmental goals aiming at GHG emission reductions is a major challenge.

Using the present situation as the reference, the environmental inventories associated with the technological options for the future were analyzed using the LCA approach (Dones et al., 1996 and 1998).

LCA Approach. LCA considers not only direct emissions from power plant construction, operation and decommissioning but also the environmental burdens associated with the entire lifetime of all relevant processes upstream and downstream within the energy chain. This includes exploration, extraction, processing, transport, as well as waste treatment and storage. The direct emissions include releases from the operation of power plants, mines and processing factories, transport systems and building machines. In addition, indirect emissions originating from materials manufacturing, and from energy inputs to all steps of the chain and from infrastructure, are covered.

Specific Characteristics of the Analysis. The previous analyses of energy systems currently in operation (Frischknecht et al., 1994), combined with information on the development trends, allowed to focus the present work on parts of the energy chains which are subject to major changes. These are: for gas systems, transport of natural gas and power plant; for coal systems, mining, coal transport and power plant; for nuclear systems,

[1] Union pour la Coordination de la Production et du Transport de l'électricité.

mining/milling, enrichment, power plant, and reprocessing; for hydro systems, power plant; and, for photovoltaic systems, manufacturing of solar cells.

For the different energy chains the most important changes in relation to the existing analysis of currently operating systems were:

- **Gas systems**: reduction of gas leakage, improvements of power plant burner performance characteristics and of overall power plant efficiency;
- **Hard coal systems**: partial methane recovery in underground mining, improvements in power plant abatement technology and of overall power plant efficiency;
- **Nuclear systems**: reductions of long-term radon emissions from mine/mill tailings, reductions of electricity consumption and CFC emissions in enrichment by replacement of diffusion by centrifuges or laser technologies, power plant improvements (particularly life time extension and increased burn-up), use of actual emissions from a modern reprocessing facility, reduced volume of conditioned radioactive solid wastes;
- **Hydro systems**: overall power plant efficiency improvements (turbine); and,
- **PV systems**[2]: improvements in the manufacturing of monocrystalline-silicon (m-Si) and amorphous-silicon (a-Si) solar cells (yield, electricity consumption), and in cell efficiencies.

The LCA methodology has been developed for and primarily applied to operating systems. Consequently, the input is normally based on the actual experience. Furthermore, the standard approach is static (i.e., only the UCPTE electricity mix at one time point was used in (Frischknecht et al., 1994) for the requirements of all processes), and its applications to future systems require extensions, extrapolations and a number of additional assumptions.

The extension of the existing data to include and characterize new technologies was based on literature, direct information from the industry and consultants, and on expert judgment. Availability of relatively detailed process information and knowledge about the relative importance of the various sources of emissions made it possible to focus the analysis and economize the use of resources. In view of the objectives of the project the parameters of primary interest were: emissions, efficiencies, material amounts (for construction and operation), and transportation requirements. Basically, the performance parameters chosen should in most cases be representative for technologies that could be implemented on a large scale between year 2010 and 2020.

The final requirements and emissions, although based on the conditions assumed specifically for the Swiss electricity supply, are not restricted by geographical boundaries. The adjustment to Swiss-specific conditions applies to the fuel origin and quality, materials production, electricity mixes, infrastructure, transportation, etc.

An example, illustrating the application of LCA to the advanced nuclear systems and the resulting radioactive emissions to air from the normal operation, is shown in Figure 3.

The environmental inventory analysis was performed on three levels: (1) individually for each system considered; (2) comparison of systems; (3) comparison of supply mix options. The evaluations of the seven options were carried out for the year 2030. This choice of year provides the upper limits that can be calculated for the yearly emissions from the whole electricity sector in the time interval considered.

---

[2] Only roof panels considered.

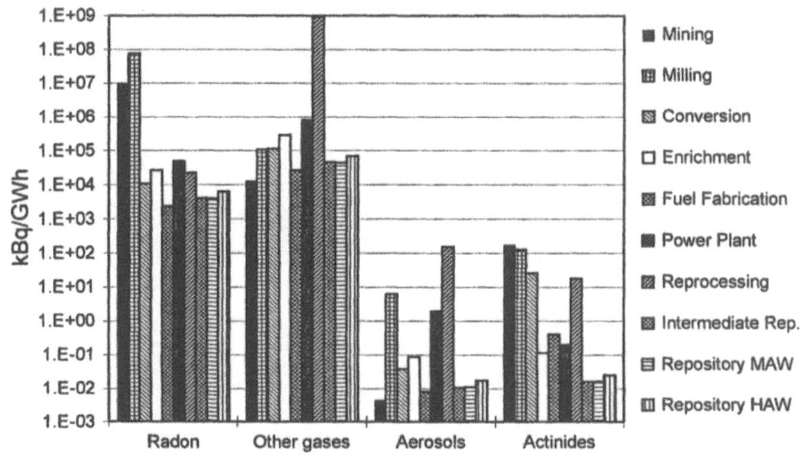

Figure 3. Radioactive emissions to air from future nuclear system (Dones et al., 1996).

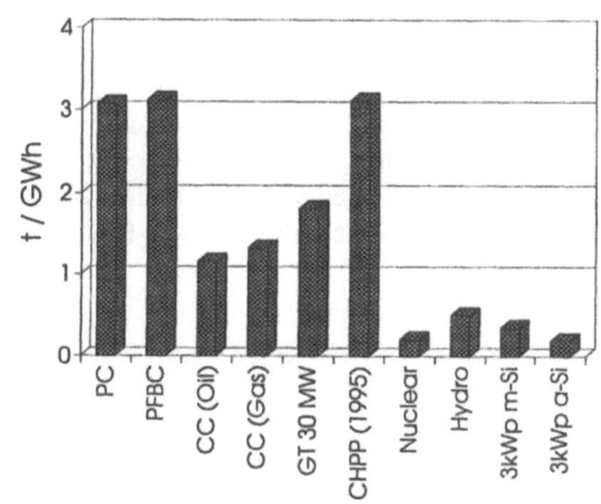

Figure 4. Steel requirements for the analyzed full energy chains (Dones et al., 1996)[4].

Systems comparisons. Figures 4 and 5 show the steel and concrete requirements, respectively, normalized by the unit of electricity. The highest steel consumption has been calculated for the hard coal systems and for the combined heat and power plant (CHPP)[3]. Nuclear and renewables are in the lower range. Coal systems rank again first with respect to the use of concrete, primarily due to the large transport requirements by rail, where large amounts of concrete are used for the infrastructure. Hydropower follows due to the great quantity of concrete used to build the dams.

[3] As a cogeneration reference technology a current gas-based 160 kWe plant was used. No extrapolation was made in this case to account for the potential future improvements of the technical and environmental performance. Consequently, the results for Combined Heat and Power Plant (CHPP) cited in this paper should be regarded as conservative.

[4] The following acronyms are used in this and some of the following figures: PC = Pulverized Coal Combustion Plant; PFBC = Pressurized Fluidized Bed Combustion Plant; CC = Combined Cycle; GT =

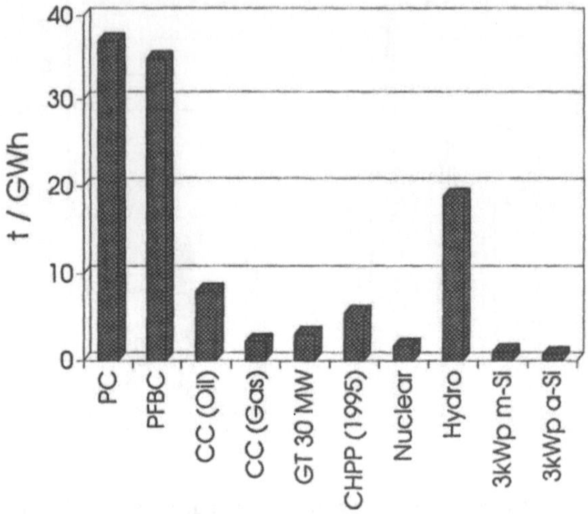

Figure 5. Concrete requirements for the analyzed full energy chains (Dones et al., 1996).

Figure 6 shows the results for the overall electricity input requirements, including contribution from all grids (Swiss and average UCPTE) at all voltages. Hydro exhibits the lowest electricity need, the photovoltaic panels with monocrystalline silicon cells the highest. For the latter systems the anticipated and here credited improvement in comparison with the currently operating solar panels is quite dramatic (reduction by a factor of about three).

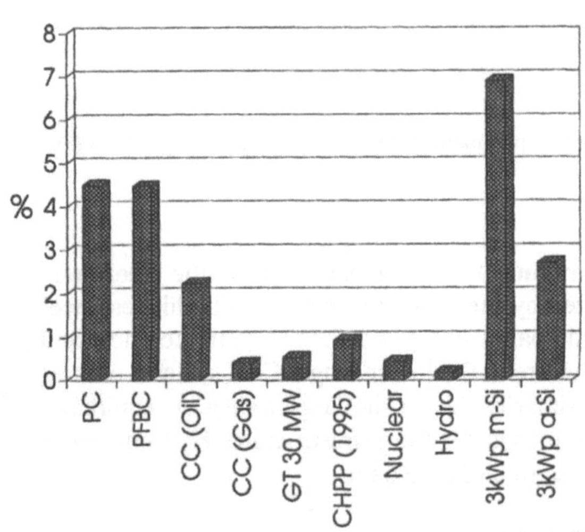

Figure 6. Electricity requirements as a fraction of the total generated by the analyzed full energy chains (Dones et al., 1996).

Gas Turbine; CHPP = Combined Heat and Power Plant; m-Si = monocrystalline silicon cells; a-Si = amorphous silicon cells.

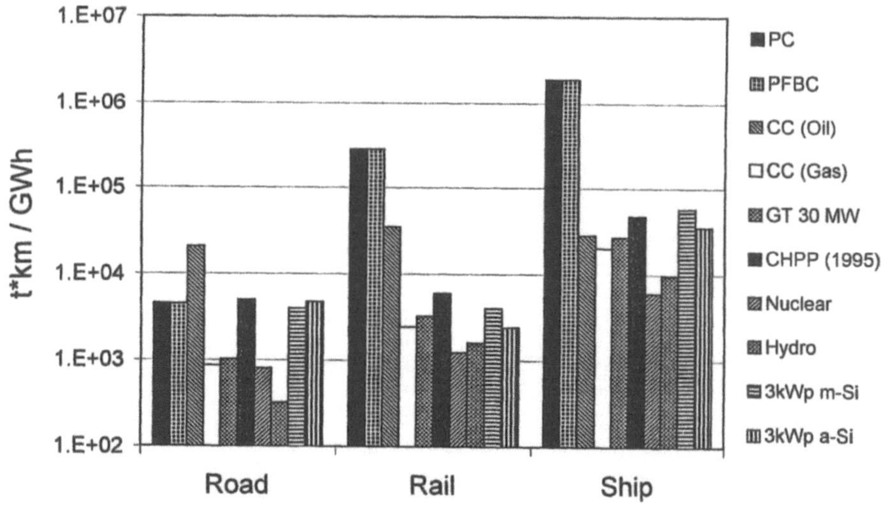

Figure 7. Total transport requirements for the analyzed full energy chains (Dones et al., 1996).

Total transport requirements are illustrated in Figure 7, divided into transport by road, rail and ship (inland waterways and transoceanic). The coal systems have by far the largest requirements of rail and ship transports, the latter up to two orders of magnitude higher than for the other systems. Hydro and nuclear full energy chains exhibit the lowest values.

In the following, only greenhouse gases (GHGs), SO_x, NO_x, and radioactivity emissions to air will be briefly discussed as examples of results obtained for the systems.

Figure 8 shows a comparison of greenhouse gas (GHG) emissions from the assessed current (average European[5] or Swiss) and future (Swiss conditions) full energy chains for electricity generation. The emissions are given in terms of grams of CO_2-equivalent per kWh electric calculated using the most recent warming potentials recommended by the Intergovernmental Panel on Climate Change (IPCC). The gases include CO_2, CH_4, N_2O, CF_4, and CFCs.

Future electricity generation systems based on fossil chains have the potential to reduce GHGs emissions by 30-50% in comparison with the corresponding present average European fossil systems. However, they remain the largest GHG producers: two orders of magnitude higher than hydro and nuclear, one order higher than PV and wind. GHG emissions from renewables and nuclear almost entirely originate from indirect sources and from upstream/downstream steps. For all systems, the largest contributor to total GHG emission is carbon dioxide. Methane emissions from coal mining and gas transport will be about 5% of total GHG for these chains.

Figures 9 and 10 show SO_x and NO_x emissions to the atmosphere, respectively. Direct emissions from the operation of power plants are given separately from all other contributions (from upstream and downstream steps of the chain and indirect). As expected, CC burning oil exhibits the greatest direct emissions of sulfur oxides at the power plant (47% of the total) as well as from the entire chain. Coal systems and gas systems have the second and third highest emissions, respectively. A PFBC power plant has very small SO_x

[5] The high value for GHG from present European gas systems is mainly due to the gas mix used, where about one quarter is blast furnace gas and coke oven gas. Would only natural gas be burnt, the total emission would be in the range 550-700 g(CO2-equivalent)/kWhe), depending on the efficiency of the power plants, the methane leakage and the energy required for pumping gas in long-distance transport via pipelines.

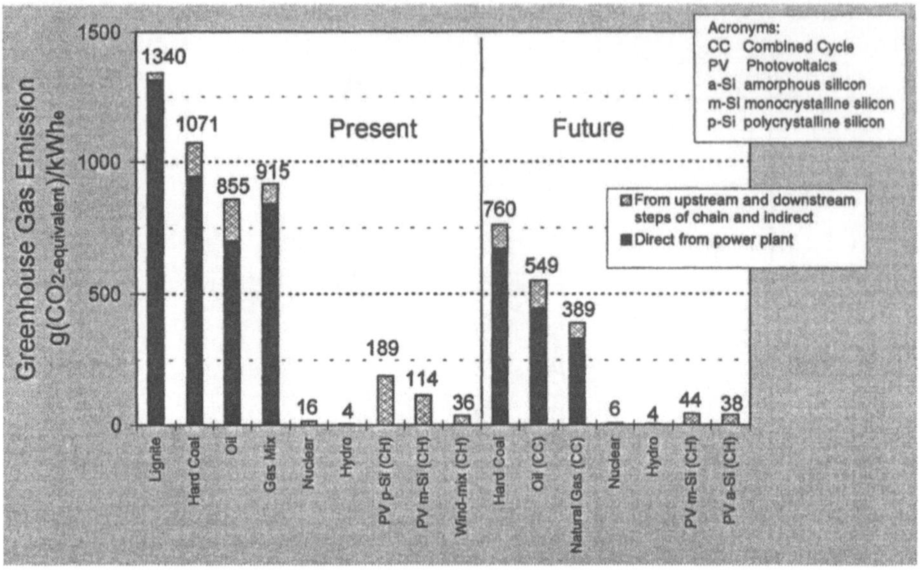

Figure 8. Greenhouse gas emissions from current and future full energy chains, using IPCC 1996 warming potentials for 100 years time horizon[6].

emissions compared to an advanced PC, thanks to its mode of operation; it will emit only approximately 8% of the calculated total from the associated full chain. The direct emissions from the gas power plant are negligible compared to the other fossil systems (2% of the total calculated for the gas chain).

As shown in Figure 10, the chain associated with PC shows the highest total emissions of nitrogen oxides (47% directly from the power plant), but the GT chain has the highest direct emissions of NO_x from the power plant. A PFBC power plant has the lowest NO_x emissions compared to the other fossil power plants (only 14% of the total from the relevant full energy chain). Among the gas systems, CC has the lowest NO_x emissions from the power plant as well as the lowest total from the chain; GT plants emit about two thirds of the total emissions from the associated chain. Oil and hard coal upstream steps show contributions to total nitrogen oxides emissions comparable with the direct ones from gas turbines.

Figures 11 and 12 show the emissions of SO_x and NO_x, respectively, divided into direct releases taking place in Switzerland (where the power plants are assumed to be located), direct external to Switzerland, and indirect ones (mostly occurring outside Switzerland). For most species, total emissions from fossil chains are primarily direct, as illustrated in the SO_x and NO_x cases.

The amount of SO_x directly emitted in Switzerland for all chains is only a small fraction of the total with the exception of PC and CC fueled with oil. For fossil systems, direct emissions of NO_x in Switzerland are comparable with the amount of other direct emissions taking place abroad, with the exception of direct NO_x releases from PFBC, which are small in proportion to the total. SO_x emissions from the nuclear and renewable energy chains are almost entirely indirect, arising mainly from electricity and steel requirements;

[6] For the present systems the results from the updated environmental inventories for energy systems (Frischknecht et al., 1996) were used.

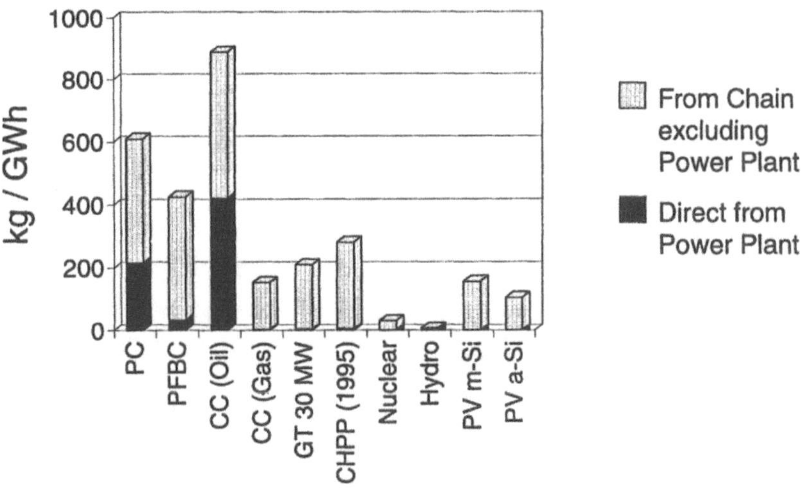

Figure 9. Contribution of power plants to total SO_x emissions to air for future systems (Dones et al., 1996).

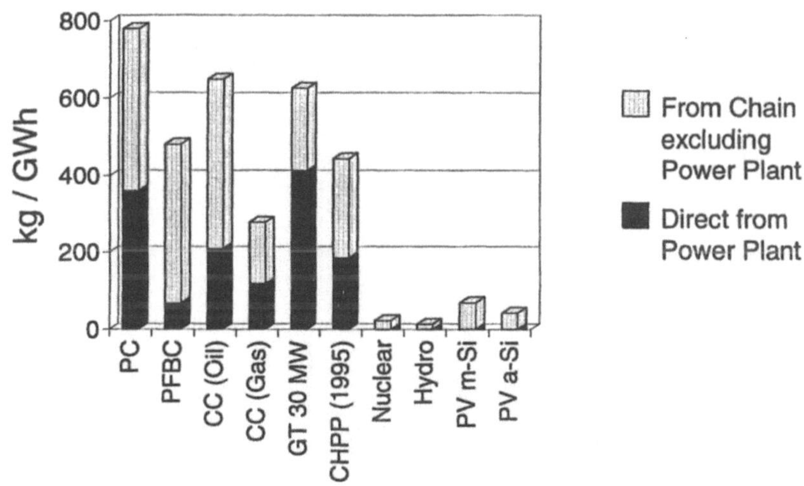

Figure 10. Contribution of power plants to total NO_x emissions to air for future systems (Dones et al., 1996).

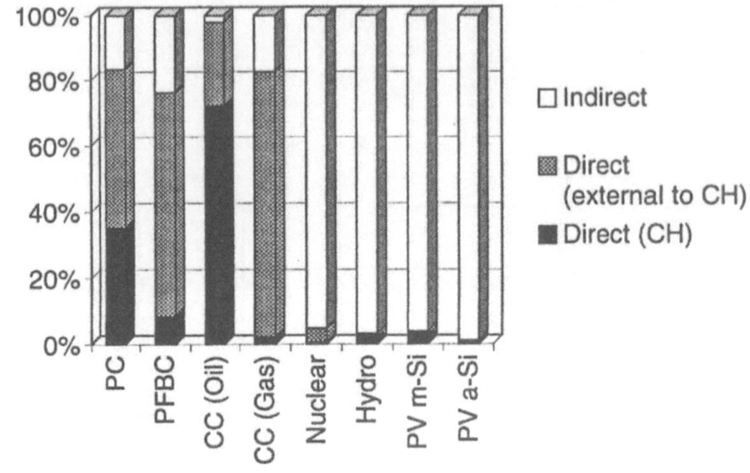

Figure 11. Direct and indirect SO_x emissions from selected future systems (Dones et al., 1996).

the same applies to NO_x emissions from photovoltaic systems. NO_x direct emissions from the nuclear chain mainly originate from processing of fuel, mining and construction machinery. NO_x direct emissions for hydro stem mostly from building machines.

A comparison of the direct radioactive emissions to air from power plants with the total calculated for the relevant chains is shown in Figure 13 for the hard coal and nuclear systems only. Results obtained for other systems are not shown because direct radioactive emissions, if any, are for these systems negligible. Four classes were chosen which include all radioactive species released to air: radon, other gases (including noble gases), actinides and aerosols.

Radioactive emissions to water and conditioned radioactive solid waste were also considered in (Dones et al., 1996) but are not discussed here; it is estimated that the quantities will decline.

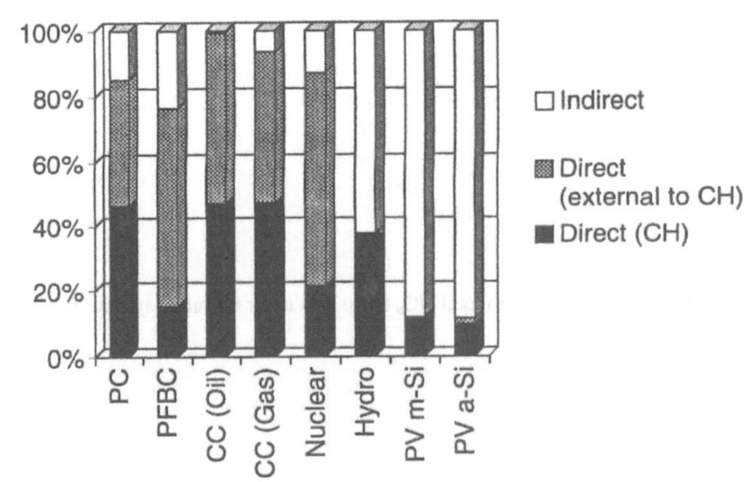

Figure 12. Direct and indirect NO_x emissions from selected future systems (Dones et al., 1996).

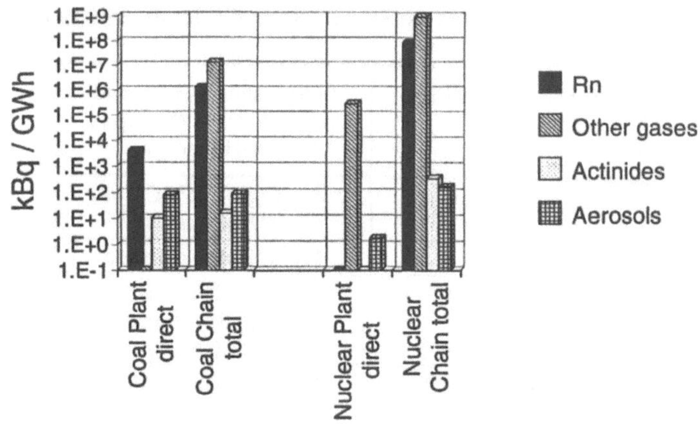

Figure 13. Comparison of the direct radioactive emissions to air from power plants and total from full chains for future hard coal and nuclear energy systems (Dones et al., 1996).

As a conclusion, the total emissions of major air pollutants are significantly reduced in comparison with the currently operating systems. Due to considerable advancements in fossil power plant technologies, the relative importance of other activities in the fossil energy chains increases. These changes are particularly important for pollutants like nitrogen and sulfur oxides.

Comparisons of supply mix options. The emissions reported in the PSI/ETHZ work (Dones et al., 1996) were used as indirect impact indicators in the study of the future electricity supply options for Switzerland (VSE, 1995). The study was conducted by the Swiss Association of Producers and Distributors of Electricity (VSE). The air emissions were assessed using the Life Cycle Assessment (LCA) approach. For details concerning the two assumed electricity demand developments (corresponding to high growth and low growth) and for the rational for the selection of supply mix options (Table 1), we refer to the VSE study. Analytical aspects, specific assumptions made and the full spectrum of results are covered in (Dones et al., 1996).

Generally, accounting for the full energy chains, the options with fossil systems show the highest releases of typical combustion gases while the options with nuclear systems present the lowest. As expected, the nuclear options exhibit the highest emissions of radioactivity. In the following, the focus is on greenhouse gases but also some other emissions are briefly discussed. With few exceptions the picture is qualitatively similar for high growth demand case options (labeled H1-H7) and for the corresponding options for the low growth demand case (L1-L7).

GHG emissions are shown in Figure 14 for both high- and low-growth demand cases. The numbering of supply mix options follows the definitions provided in Table 1. The "fossil" option H1 exhibits the highest total yearly emission of GHGs that originate from the entire chains of the various systems included. A total GHG emission of 26.2 Mt(CO_2-eq.)/yr. is calculated for H1, of which 84% is directly produced by the power plants; 92% of the total is from the systems covering the gap. The "nuclear" option H2 shows the lowest total yearly production of greenhouse gases from the relevant whole chains. About 2.3 Mt(CO_2-eq.)/yr. are produced in this option, merely 12% thereof are from the nuclear energy chain covering the gap. 2.0 Mt(CO_2-eq.)/yr. are emitted from the base supply alone. The other options have intermediate values in the range 9.2 to 24.3 Mt(CO_2-eq.)/yr. (or 35% to 93% of the value calculated for H1).

Table 1. VSE supply mix options to cover the future electricity gap.

Supply mix options		System shares in the cold season to cover the gap of electric energy
1	Conventional thermal	50% CC(gas); 25% CC(oil); 25% Coal
2	Nuclear	100% Nuclear domestic
3	Mix Nuclear/Gas	Today's installed nuclear capacity; rest of supply CC(gas)
4	Import 100%	10% CC(gas); 30% Coal; 60% Nuclear
5	Import 50%	Import: 10% Coal; 40% Nuclear. Domestic: 25% CC(gas); 25% Nuclear.
6	Conventional thermal + PV	as Option 1 with PV substituting 5%
7	Mix Nuclear/Gas + PV	as Option 3 with PV substituting 5%

CC = Combined Cycle; PV = Photovoltaic (only roof panels considered)

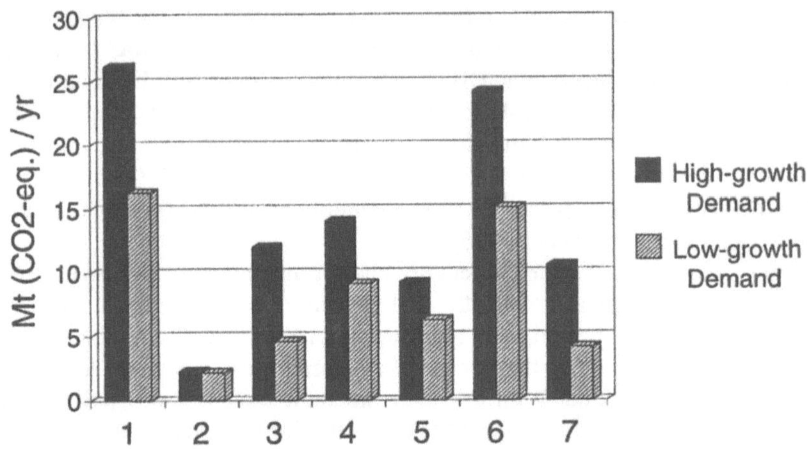

Figure 14. Greenhouse gas emissions for the 7 supply mix options of the high and low growth demand cases in year 2030 (Dones et al., 1996).

To put in perspective the calculated total GHG emissions from the electricity sector, the various supply mix options can be compared with the corresponding present inventories for Switzerland. The total GHG emission from the current Swiss electricity mix, calculated using LCA in (Frischknecht et al., 1996), is nearly 1.0 Mt(CO_2-eq.)/yr., of which about 63% is from the energy chain associated with the oil power plant in Vouvry, contributing typically only about 1.2% to the total electricity generated.

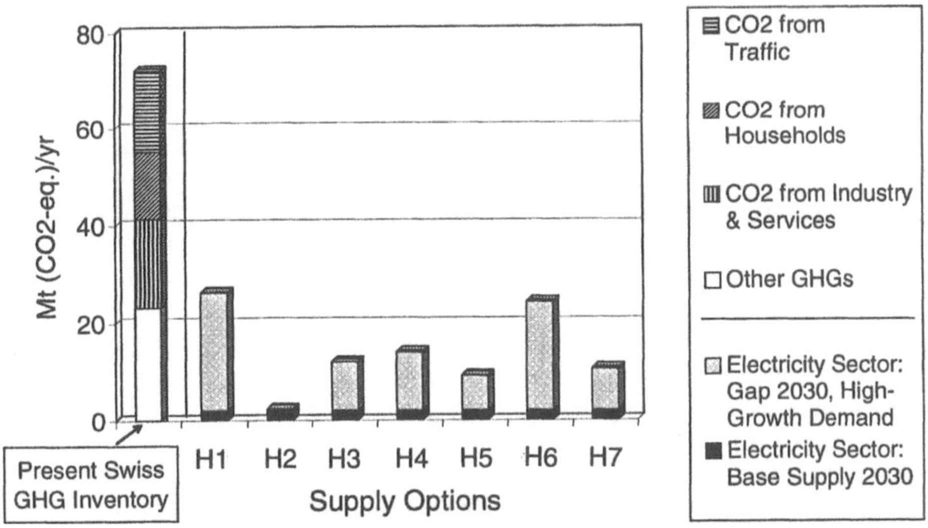

Figure 15. Comparison of the present Swiss GHG emission inventory with LCA results for high-growth demand case options for year 2030 (Dones et al., 1996).

Figure 15 shows a comparison of the present Swiss GHG emission inventory with LCA results for the supply mix options for the high growth demand case in year 2030. The total emission inventory of GHG actually within Switzerland was estimated[7] as 72.2 Mt(CO_2-eq.)/yr. The highest share of the present inventory is from carbon dioxide, 68% or 49.1 Mt(CO_2)/yr. The total amount of GHGs that would be emitted from the electricity sector (including full chains, inside and outside Switzerland) in the case of the "fossil" option H1 in 2030 corresponds to about 36% of the current total Swiss domestic GHG inventory. The corresponding share for option L1 is 22%.

Similar patterns apply to the other combustion products such as SO_x, NO_x, NMVOCs and particulate matter. H1 (L1) supply mix options exhibit the highest total yearly production of these pollutants and H2 (L2) the lowest. SO_x emissions calculated for H1 would approximately equal 34% of the present total inventory in Switzerland (about 21% for L1). For the NO_x the corresponding numbers are 17% for H1 and 11% for L1. For comparison, the contributions of the current electricity mix to the total emissions are about 6% in the case of SO_x and less than 2% for NO_x.

With regard to radioactivity the nuclear options exhibit the highest emissions and the purely fossil ones the lowest. The contribution of the nuclear power plant to the overall emissions from the full chain is very small (less than 0.1%). In comparison with the current situation there is nevertheless a reduction of the total estimated radioactive emissions due to the specific improvements in the nuclear chain. The radiological burden associated with these emissions is lower than 1% of that resulting from the natural background and medical treatment.

[7] This is based on (BUWAL, 1994a and 1994b) after correcting for the GWP according to IPCC 1994 (Houghton et al., 1995).

Perspective on Environmental Impacts

Environmental Inventories as Impacts Indicators. It is possible to use LCA as a tool for simplified impact assessment. Pollutants can be aggregated into, for example, 13 environmental impact classes such as: greenhouse effect, ozone depletion, acidification, photosmog, nitrification, radioactivity. Impact analysis based on LCA is, however, subject to considerable simplifications and the results exhibit the corresponding limitations (Hirschberg, 1995). The LCA approach does not distinguish between the physical characteristics of the emissions (e.g. rate, duration, location), meteorological conditions, complex pollutant interactions and transformations. Consequently, for some categories such as photosmog the results may not be always meaningful due to the dependencies and nonlinearities involved. For some other impact classes, LCA-based impact estimation may represent a more valid and resource-saving approach, complementary to the impact pathway method.

In (VSE, 1995) LCA-based emissions were used as a surrogate for impacts. With the exception of the greenhouse effect no aggregation into impact classes was made. For the greenhouse effect, which represents a global impact, the place and time characteristics of the emissions are practically of no importance. Furthermore, the estimates of the real impact of global warming are associated with enormous and partially uncontrolled uncertainties. For these reasons the emission-based results are considered as fully relevant as an impact indicator. For other pollutants and impacts the limitations of the corresponding simplification are more serious. In this context the LCA-based emissions should be regarded as an indicator of possible impacts and in no way as a measure of expected impacts. A relative ranking of such impacts can be undertaken using in addition to the estimated inventories the experience from EC/US studies on external costs of fuel cycles (EC, 1995; ORNL/RfF, 1992 - 1995).

Some Relevant Experiences from State-of-the-art External Cost Studies. The state-of-the-art external cost studies consider a full range of impacts and employ the impact pathway, or damage function, approach. The focus of the analysis is on the most significant discharges and impacts. In a bottom-up analysis, individual pathways are identified to link the source of the emissions, the dispersion of these emissions, their incremental human health and environmental impacts, and, if desirable, the economic value of these damages. Because of its more specific, detailed nature, the bottom-up approach is able to account for site-specific differences that affect the nature and the extent of human health and environmental impacts. This site-specific aspect is also a limitation of the approach. The single plant focus of the studies makes it more difficult to issue generalized statements of environmental impacts for the purpose of, for example, national energy policy making (Lee et al., 1997).

It should be noted that not all methodological issues have been resolved. New insights concerning key contributors are continuously being gained and have not been generally implemented in all studies. According to the experience[8], environmental and health impacts for technologies that meet stringent regulation may be significant but are not overwhelming. The exact magnitude of externalities varies depending on type and amount of emissions, population distribution, meteorological conditions, valued ecosystems, treatment of severe accidents, boundaries of time and space, degree of internalization etc. Climate change,

[8] The primary (but not the only) basis for the summary of experience provided here are the EC and US studies of external costs of fuel cycles (EC, 1995; ORNL, 1992 - 1995). The results of these studies are, however, subject to modifications due to recent findings and extensions. These insights, which partially have not yet been published, are also reflected in the present survey.

public perception of the nuclear option and nuclear wastes storage remain as controversial issues.

Excluding global warming (which may lead to environmental and health impacts), health effects dominate over the environmental ones. This picture is amplified by the recent findings concerning chronic and acute mortality due to particulates. Table 2 shows an overview of the impacts that seem to dominate within each energy chain. However, it should be noted that there are large difference between the absolute results obtained for the different chains. Furthermore, the actual ranking of the dominant contributors is technology and location dependent.

Impacts of future supply mix options. The impacts are here discussed in view of the LCA-based burdens and findings of the major external costs studies. Given the present state of knowledge, the evaluation cannot at this stage be absolute. Instead a relative ranking of impacts can be made in relation to the current electricity supply mix. Generally, one needs to distinguish between the two cases of high and low growth when discussing impacts. The

Table 2. Most important impacts.

Energy Chain	Impact
Coal	Climate change SO_2 and NO_x via aerosols (acute mortality) Particulates (chronic and acute mortality) Ozone (health, crops) Transportation and mining accidents Truck road damage
Oil	Climate change SO_2 and NO_x via aerosols (acute mortality) Particulates (chronic and acute mortality) Ozone (health, crops) Oil spills
Gas	Climate change NO_x via aerosols (acute mortality) Particulates (chronic and acute mortality) Ozone (health, crops) Pipeline leakage and accidents (regional)
Nuclear	Long-term impacts of C-14 from reprocessing Long-term impacts of Rn-222 from closed mill tailings Occupational health Reactor accidents
Biomass	Particulates (chronic and acute mortality) Transportation impacts
Hydro	Local ecosystem impacts Visual amenity Recreational impacts Large dam accidents
Wind	Visual amenity Noise if in populated areas Occupational accidents

reason is that for some of the options the share of the fossil systems remains on a low level in the low growth case.

For global impacts the total emissions can be considered as representative for the impacts. According to the experience, regional impacts dominate in most cases over the local ones. Health impacts which in turn dominate over environmental effects are not explicitly considered in the VSE study (VSE, 1995). The issue of severe accidents was outside of the scope of the study which was limited to normal operation. As discussed before use of burdens alone as a surrogate for impacts is subject to serious limitations. Here we provide a preliminary, qualitative ranking of the possible environmental and health impacts of VSE's supply options, taking into account the findings in (Dones et al., 1996), and considering simultaneously the insights from the recent external cost studies. The evaluation, which is limited to the normal operation and air pollutant emissions, is carried out using the current Swiss mix as the reference. The results are shown in Table 3. Apart from the "Nuclear" case, options "Mix", "Import50" and "Mix&Solar" exhibit the best performance with regard to the environmental and health criteria.

Current Developments. The simplified impact assessment presented here has elements of improvisation; while the approach is supported by numerical results the

Table 3. Qualitative evaluation of the impacts of supply mix options (only normal operation considered)

a) Local, regional and global environmental and health impacts; low growth case.

Option→ / Type of impact↓	1 Fossil	2 Nuclear	3 Mix	4 Import 100%	5 Import 50%	6 Fossil &Solar	7 Mix &Solar
Local&regional; Health&environmental	considerable increase	decrease or very small increase	decrease or very small increase	moderate increase	decrease or very small increase	considerable increase	decrease or very small increase
Global; Health&environmental	large increase	decrease or very small increase	moderate increase	considerable increase	moderate increase	large increase	moderate increase

Option→ / Type of impact↓	1 Fossil	2 Nuclear	3 Mix	4 Import 100%	5 Import 50%	6 Fossil &Solar	7 Mix &Solar
Local&regional; Health&environmental	large increase	decrease or very small increase	moderate increase	considerable increase	moderate increase	large increase	moderate increase
Global; Health&environmental	large increase	decrease or very small increase	considerable increase	considerable increase	considerable increase	large increase	considerable increase

decrease or very small increase
moderate increase
considerable increase
large increase

Note to the tables: The possible impacts are scaled relative to the health and environmental impacts of the current Swiss electricity sector.

absolute impacts cannot be fully quantified at this stage. However, the full scope environmental impact analysis is presently carried out by using a mesoscale dispersion model for the simulation of transport and chemistry of airborne pollutants (Andreani et al., 1995; Andreani and Keller, 1996). This work will ultimately allow the full simulation of air pollution impacts associated with various energy scenarios for Switzerland. Emission scenarios are presently being evaluated for NO_x and VOCs. Preliminary simulations of maximum ozone concentrations have been carried out. The dispersion model will be linked with environmental and health impact models to enable the calculation of total impacts. Figure 16 shows the overall scheme for the environmental impact assessment as applied in the GaBE project.

Severe Accident Risks

Background. Treatment of severe accidents has been considered for a long time as one of the unresolved topics in comparative assessment of energy systems. This is primarily due to the following reasons:

- Technical comparison of risks is subject to a number of inherent and practical limitations. In the past, the completeness and consistency of the data available for comparison has been clearly unsatisfactory.
- In the context of decision-making the technical perspective on risks (normally expressed in quantitative terms) is one among other factors of importance. Public perception of risks, influenced by sociological and psychological aspects, had and continues to have a major impact on decisions.

Here we focus on the technical comparison of severe accident risks. While the substantial influence of subjective aversion towards the possibility of very high damages/losses on the behavior of individuals is acknowledged, informed decisions should

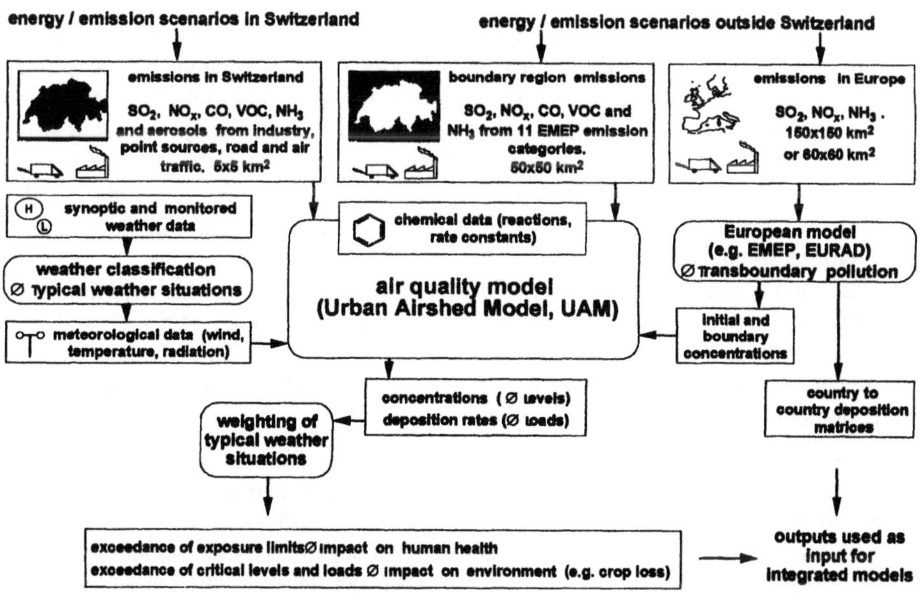

Figure 16. Environmental impact assessment within the GaBE project.

be taken in full knowledge of the experts' estimates of the likelihood of accidents. In (Lee et al., 1997) it is pointed out the two perspectives on risks (technical and perception-based) are complementary and should not be considered as alternative views.

With respect to the technical risk evaluation, significant progress has been achieved in recent years, enabling carrying out a better founded comparative risk assessment than was the case in the past. In particular, the set of available data has been considerably extended and their quality (i.e. completeness and consistency) has been greatly improved. Furthermore, while in the past all numerical comparisons focused on fatalities, a more inclusive spectrum of health, environmental and economic indicators related to severe accidents can now be addressed. In spite of these specific improvements there remains potential for further advancements with respect to dealing with some of the practical limitations.

Analytical Options and Analysis Scope. Comparative analysis of severe accidents can be based on historical evidence, on Probabilistic Safety Assessment (PSA)[9], or on combinations of these. The need of introducing the probabilistic approach stems from some of the basic limitations of the retrospective analysis:

- In some cases there exists quite weak statistical evidence and very limited representation of the full spectrum of hypothetical accidents. A complicating factor is the heterogeneity of the available data in terms of technologies as well as with regard to the operational and physical environments in which they operate.
- Depending on the purpose (and object) of comparative analysis the experience-based data may not be applicable. Given large contrasts between the safety-related characteristics of the systems represented in the databases and systems being subject of the analysis, historical data may be questionable or even irrelevant[10].

With respect to the scope of a comprehensive comparative analysis of severe accidents some consensus has been reached (Hirschberg et al.,1998; IAEA, 1992):

- The comparison should not be made on the basis of consequences of severe accidents in isolation. Also the associated frequencies must be estimated.
- The comparison should cover all the relevant steps in full energy chains, i.e. whenever applicable apart from power production (conversion) also exploration, extraction, transports, processing, storage and waste disposal. In fact, for some energy chains (applies to fossil cycles) these other steps may represent a larger hazard than the power plant itself. If feasible, following the "cradle to grave" principles of Life Cycle Assessment (LCA) also the risks associated with the different life phases (i.e. construction, operation and dismantling) of the relevant facilities should be addressed.
- Time and space dimensions of accident consequences are of interest. Recent state-of-the-art external cost studies attempt to consider local, regional and global impacts in short, medium and long term perspective. This applies to effects associated with normal operation as well as to severe accidents that have been considered (primarily nuclear).

---

[9] This term is generally used within the nuclear industry. A corresponding type of analysis employed within the process industry is usually referred to as Quantitative Risk Analysis (QRA).

[10] This situation may occur for example in strategical studies of electricity supply options for the future. The lack of relevance of historical data is amplified whenever the future options represent major safety-related advancements and at the same time the historical data are dominated by events at facilities which are far from current safety standards.

- In order to carry out the comparison, representative indicators need to be established. These indicators should not be limited to the number of fatalities but should ideally reflect various types of consequences of severe accidents, such as: number of fatalities, with a distinction between immediate and delayed fatalities; number of injured; number of evacuees and, if feasible, duration of evacuation; extent of ban on consumption of food; polluting release of hydrocarbons or other toxic chemicals; land or water area being subject to enforced clean-up, or being interdicted/condemned; economic loss.

Given the availability of these indicators comparison may be carried out by normalization based on energy outputs specific for each energy carrier.

It should be noted that some important aspects of severe accidents are not amenable to straight-forward quantification. This applies in particular to environmental effects such as loss of quality, aesthetic values, disturbance of the ecosystem or genetic deterioration, possible irreversibility of damages, and to social impacts of psychological nature. This calls for qualitative coverage.

Use of Historical Data. In order to identify severe accidents using historical records it is necessary to define the specific attributes of a severe accident. This definition needs to include the various consequence (damage) types and a minimum level for each damage[11]. In the literature there are major differences between the various severe accident definitions that have been used. This applies both to the set of specific consequence types considered and the damage quantity regarded as severe. Generally, the criteria are arbitrary, not standardized and can change with time.

Table 4 displays the various definitions used. The definition formulated and implemented in Paul Scherrer Institute's database ENSAD (Energy-related Severe Accident Database; Hirschberg et al., 1998) is the most inclusive one in terms of damage categories covered. There are many parallels between this definition and the one proposed for future implementations in (IAEA, 1992), although ENSAD uses lower threshold for the number of fatalities and for economic consequences.

The ENSAD database (Hirschberg et al., 1997 and 1998) was established using nine major commercial and non-commercial accident databases, and numerous additional sources of information (annual publications, general and specialized literature, national and international newspapers, incident lists and reports, direct contacts with responsible companies and other competent organizations or individuals) to enhance its completeness and quality. The database covers full fossil (coal, oil, gas), nuclear and hydro energy chains. It has been demonstrated that ENSAD, thanks to using a variety of inputs, has a superior world-wide coverage of energy-related accidents compared to all other databases considered[12].

Currently, ENSAD contains 3319 energy-related accidents, with nearly 86.7% occurring within the time period 1969 - 1992, the period used for comparative

[11]Accidents which fall under this minimum level should not be ignored in a comparative assessment. They should be accounted for along with the impacts from normal operation. Treatment of small accidents is, however, not controversial; they are not subject to aversion and the complex issue of estimating frequency of catastrophic events does not enter the discussion in this case.

[12]This applies generally to severe man-made accidents represented in ENSAD, at least with regard to person-related consequences (fatalities, injuries, evacuations). Nevertheless, the non-energy-related severe accidents were not subject to complementary searches and checks and, consequently, the quality of these data is lower.

Table 4. Different definitions of severe accidents (Hirschberg et al.,1998).

Damage Categories (OR-mode)	Sources					
	DNV (1992)	Glickman and Terry (1994)	Swiss Re (1994)	Rowe (1977)	IAEA (1992)	ENSAD (Hirschberg et al., 1997)
Fatalities	≥ 1	≥ 5	≥ 20	-	≥ 10	≥ 5
Injured	-	-	≥ 50	-	≥ 10	≥ 10
Evacuees	-	-	-	-	≥ 200	≥ 200
Homeless	-	-	2000	-	-	-
Ban on consumption of food	-	-	-	-	yes	yes
Polluting release of hydrocarbons and chemicals[a]	≥10,000 tones	Yes (no minimum specified)	-	-	-	>10,000 tones
Enforced clean-up of land+water	-	-	-	-	≥ 25 km^2	≥ 25 km^2
Economic loss (million $US\$_{1992}$)	≥ 2	-	≥ 58.6 (total loss) Insured loss[b]: Shipping: ≥11.7 Aviation: ≥23.4 Other: ≥29.3	≥ 6.9	> 10	≥ 5

[a] The release levels refer exclusively to hydrocarbons. Other chemicals need to be treated on case-by-case basis.

[b] The lower limits for the insured damages are increased annually by the inflation rate.

evaluations[13]. Applying the definition of severe accident, as established in ENSAD, 1772 severe energy-related accidents were identified. Furthermore, 807 of the energy-related accidents caused five or more fatalities, with 636 occurring during 1969 - 1992. For the other consequence categories the number numbers of energy-related accidents in ENSAD are as follows[14]:

≥10 Injured:	472
≥200 Evacuees:	146
≥10 000 tones of pollutive hydrocarbon releases:	607
≥25 km^2 area of enforced clean up (land + water):	30
≥5 millions 1992 US$ of economic loss:	292

For specific energy chains the number of accidents with five or more fatalities is according to ENSAD for the period 1969 - 1992 as follows:

Coal:	133
Oil:	295
Natural gas:	88
LPG:	77
Nuclear:	1
Hydro:	13

[13] The data in ENSAD go much further back in time. The completeness of very old data is however problematic. The lower limit for evaluations was chosen with view to the temporal changes.

[14] Adding the number of accidents within each damage category (including accidents with ≥5 fatalities) exceeds the total number of severe accidents (1772). This is due to the fact that some accidents result in a variety of damages and are therefore accounted for within several categories.

Table 5. Comparison of severe accidents associated with various energy chains; normalized, energy-related severe accident records for the period 1969 - 1992, world-wide and for the hydro and nuclear chains also aggregated for OECD; immediate fatalities, injured, evacuees and economic losses per unit of energy (Hirschberg, 1997 and 1998).

	Coal	Oil	Natural Gas	LPG	Hydro		Nuclear	
					World-wide	OECD	World-wide	OECD
Immediate Fatalities per GWe*a	3.12E-01	4.13E-01	9.70E-02	4.0	8.30E-01	1.62E-02	1.18E-02	0.0
Injured per GWe*a	2.34E-02	4.35E-01	2.55E-01	17.54	6.83E-03	-	1.41E-01	0.0
Evacuees per GWe*a	-	7.65	7.49	665.89	36.23	11.28	106.48	74.71
Economic losses in millions 1992 US$ per GWe*a	7.02E-03	6.35E-01	8.83E-02	2.74	6.40E-01	8.15E-01	124.09	2.66

Using this database evaluations of severe accident frequencies and their consequences were carried out for each energy carrier covered in (Hirschberg et al., 1997 and 1998). These results were then used for comparisons between the various energy sources. The comparisons concern the electricity sector, although within the gas chain also the Liquid Petroleum Gas (LPG) was included. The results were normalized on the basis of energy production by means of each of the sources covered. The data on the various energy carriers, used for normalization, origin primarily from the International Energy Agency (IEA, 1993). Table 5 provides the results of the comparison for four indicators which can be compared over all energy chains. Other indicators (e.g. large release of pollutive hydrocarbons) are not applicable to all chains. For the nuclear and hydro chains, where risks are concentrated to the plants generating electricity, a distinction is made in the table between the results obtained using world-wide statistics on the one hand and considering OECD countries only, on the other hand. The rationale for this separation will be provided in the following; also nuclear-specific results obtained for latent fatalities will be discussed below.

Some comments on the completeness and quality of the numerical indicators presented in Table 5 are in place.

1. As opposed to most previous major studies the comparison reflected in the table covers a number of categories beyond fatalities, including injuries, evacuations, and economic damages.
2. Clearly, for the damage categories other than fatalities the completeness of the data is more of a problem and the associated uncertainties are substantial.

3. The comparability is affected by the fact that the quality of the material is not homogeneous over all energy chains. A weak point in the context of immediate fatalities is the data on coal mining accidents in China. Latent fatalities are commented on below. Generally, records on injuries in coal and hydro chains are very scarce while they are much more satisfactory for the other chains. The records concerning evacuations are quite poor for the hydro chain where at the same time evacuations are a very important aspect of large accidents. At this stage the time factor is not explicitly represented when comparing the evacuation records. The duration of evacuation may range from hours to practically infinite time[15], which needs to be taken into account when judging in this context the severity of an accident. Economic losses constitute the most uncertain indicator, deserving a separate comment.
4. Latent fatalities are particularly relevant for nuclear power as manifested by the Chernobyl accident[16]. For other energy chains the potential delayed fatalities due to severe accidents are of a different nature and not possible to estimate based on the current state of knowledge but are at the same time expected to be of much lower importance. For some large accidents in the fossil and hydro chains there are indications that deaths occurring with a certain delay, but still connected with these accidents, may not have been included in the records (i.e. were not allocated to these accidents).
5. The comparison of economic damages is limited by incompleteness and some serious inconsistencies. First, the estimates of monetary losses are not available for a major part of non-nuclear accidents. Second, the cost elements covered, i.e. the boundaries of the assessment, are normally not documented and may vary widely from case to case. Third, the nature of the reported costs may be different - there is normally a large discrepancy between the compensation paid by insurance companies, claimed damages, real damages, direct costs and indirect costs. Thus, the results of the comparison and their interpretation are subject to serious reservations.
6. In spite of the higher uncertainties associated with indicators other than immediate fatalities the comparisons are meaningful and will hopefully stimulate further efforts aiming at reducing the degree of incompleteness and the number of inconsistencies.
7. Some types of damages are presently not covered in comparisons since the basis for comparison is weak or they are not applicable to all the energy chains. Examples include oil spills or land contamination. For the treatment of these impacts analyses of specific chains should be consulted (Hirschberg et al.,1998).

In Figure 17 immediate fatalities for various energy chains are displayed in the form of normalized frequency-consequence curves. The figure illustrates the catastrophic potential of the various energy carriers and some distinct differences between them. Coal and oil chains exhibit the highest frequency of accidents up to the level of about 70 fatalities per GWe*a while hydro has the lowest. The reverse applies at higher levels of consequence. Latent fatalities, particularly relevant for the nuclear chain (or more specifically for the Chernobyl accident), need to be considered in order to obtain a more complete picture.

[15] This applies particularly to severe hydro and nuclear accidents. Due to the associated long-term land loss there may be a need to relocate a large amount of people. Thus, in the case of the Chernobyl accident 210 000 people were subject to relocation, either through enforced measures or spontaneously; the social disruption accompanying this process is possibly the most serious consequence of this accident.

[16] The current best estimate of the Chernobyl-specific delayed fatalities, primarily based on the assessed doses received by the emergency workers and by the public, is in the range 2.9 to 10.4 fatalities per GWe*a produced by the nuclear energy world-wide (Hirschberg et al., 1997); the upper bound was obtained using no exposure threshold, an approach not recommended by the Health Physics Society (Moosman et al., 1996).

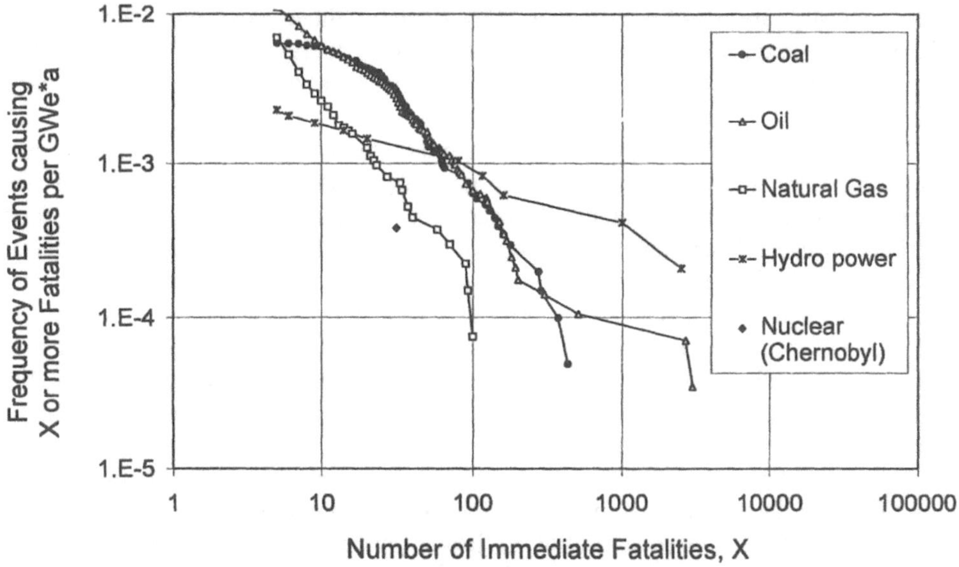

Figure 17. Frequency-consequence (immediate fatalities) curves for different energy chains, based on historical accidents world-wide in the period 1969 - 1992 (Hirschberg et al., 1997 and 1998).

Apart from the problems with temporal changes (technology, safety regulations, safety culture, efficiency of emergency services etc.), the use of historical data provides only a partial measure of the risks posed by the different hazards. Events which could, but have not yet occurred, are inevitably excluded. The presentation of aggregated world-wide data does not give information on the variability between countries, regions, technologies, systems, operational standards, safety cultures and sites. Homogeneity in both time and geographic space is implicitly assumed. For some technologies the lack of homogeneity is so pronounced that the validity of aggregated, generic estimates is very limited. For this reason, when presenting the results for nuclear and hydro chains in Table 5 a distinction has been made between world-wide statistics on the one hand and results obtained for OECD countries on the other hand.

In order to explicitly represent a specific technology operating at a specific site and within a specific operational environment, Probabilistic Safety Assessment (PSA) or Quantitative Risk Analysis (QRA) needs to be employed.

Predictive Approach. The PSA-based approach is prospective in nature as opposed to the retrospective analysis based on historical data. A PSA or a QRA provides a logical model integrating the response of the plant systems and operators to the internal and external accident initiators. The model represents a decomposition of the system to the (component) level, which enables the use of component reliability data based on the operating experience. Subject to suitable modifications and frequently under the name of QRA, PSA has also been adopted within the space, offshore and process industries. However, this has been subject to substantial country-to-country variation and to a much more narrow range of applications. This stems partially from the fundamental differences between the nuclear industry (where the highest risks are associated with essentially one process) and the other ones (particularly within the chemical industry a multitude of interdependent processes may be at work within the same plant). Some facilities within the various fossil energy chains exhibit similarities with the process industry. In a number of countries (particularly the Netherlands, UK and USA) the QRA applications within the process industry are steadily growing. Nevertheless, there exist only few published applications, relevant for the comparative assessment of energy systems. Another problem is the difference in scope and depth of applications as practiced within the various industries.

The PSA perspective is particularly important for energy chains where the severe accident risks are dominated by electricity generating plants (point sources) and where historical evidence is either statistically scarce or its applicability is subject to serious limitations. Western hydro and nuclear power plants satisfy all these conditions. On the other hand, within the coal energy chain the accidents are concentrated to the mining stage; within the gas and oil chains a vast majority of accidents occurs in the transportation stage under a wide spectrum of conditions. These activities do not lend themselves to the application of probabilistic approaches to the same extent as centralized facilities. Moreover, for these systems the historical accident experience is extensive and its applicability much less of a problem.

Figure 18 shows the PSA-based normalized frequency-consequence curves for the damage category latent fatalities to 800 km for the Swiss nuclear power plant Muehleberg (Cazzoli et al., 1993). Different confidence levels (5%, 50% and 95%) are shown in the figure; the mean value can be regarded as the main reference, while the 95-th percentile can be interpreted as providing a bounding value. No credible accident scenarios leading to immediate fatalities among the public have been identified for Muehleberg. Along with the PSA-based curves a range of values for predicted Chernobyl-specific latent fatalities is shown in the figure; this range of values reflects the impact of using a dose cut-off. The large difference between the PSA-based results obtained for Muehleberg and the experience-based Chernobyl results illustrates the limitations in applicability of past accident data to cases that are radically different in terms of technology and operational environment.[17]

Comparing with Figure 17 the above PSA-based Muehleberg-specific frequency-consequence (mean value) curve for latent fatalities is typically at the frequency level by two orders of magnitude lower than the generic experience-based frequency curves for immediate fatalities within coal, oil, natural gas and hydro energy chains. For the extreme consequences involving several thousands of fatalities this difference becomes even larger (in relation to oil and hydro chains; the other chains do not exhibit any historical events at this level of consequence). Few comments are here in place:

- The Muehleberg plant though relatively old has been extensively backfitted to current high safety standards. Furthermore, the early off-site risks are negligible in this case due to relatively low radionuclide inventory and low population density in the immediate proximity of the plant. The results obtained for this particular case probably represent the lower range of consequences for currently operating nuclear power plants.
- Accident management measures have been integrated in the Muehleberg analysis, as is the current practice in nuclear PSAs. For the emergency countermeasures, conditions best approximating the current Swiss off-site protective action strategies have been modeled.
- Damage categories other than fatalities are of high interest and are reflected for example in (Cazzoli et al., 1993). Of particular importance is land contamination, manifested by the estimated condemned and interdicted land areas. For the most

[17] In (Hirschberg and Cazzoli, 1994) an extended PSA was used for the estimation of external costs associated with hypothetical nuclear accidents. This approach was contrasted with the use of Chernobyl-specific results as the reference in the same context.

severe accidents these areas may be large, though at extremely low frequency levels.

ANALYSIS INTEGRATION

As outlined in the preceding sections the approaches used in the analysis of energy systems include:

- Life Cycle Assessment (LCA), primarily for the detailed analysis of environmental inventories; the results cover gaseous and liquid emissions, as well as energetic and non-energetic resources (e.g. raw materials).
- Environmental and health impact assessment based on simulation of the transport, chemical conversion and deposition of major pollutants.
- Direct use of historical accident experience and applications of probabilistic safety assessment for analysis of the severe accident potential of the systems of interest.
- Modeling of the economic impacts of environmental policies through application of a large scale "bottom-up" energy-economy model.

In addition, relevant insights from recent major external cost studies flow into the overall assessment. Given the complexity of the problem as well as the diversity of the indicators and interests, the way towards decision-making is long and involved. Apart from technical, ecological and economic aspects, social preferences need to be considered. Multi-criteria analysis provides a framework that allows the often conflicting socioeconomic and ecological criteria to be addressed simultaneously. Figure 19 illustrates the principles for analysis integration as applied within the GaBE project. This includes the interactions and influences affecting the various analytical modules. All links indicated in the figure have not been implemented yet. Some of these links are of qualitative nature and do not correspond to data flowing from one module to another.

The tools and full scope application of multi-criteria analysis are under development. A pilot application was carried out using VISA, a computerized aid to multiple criteria

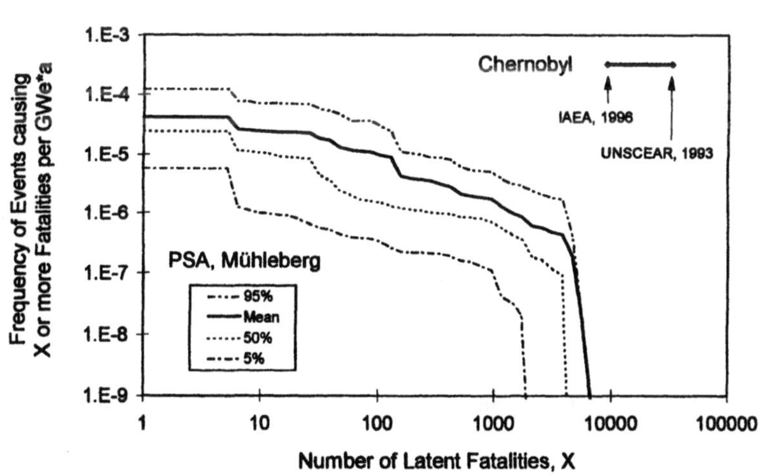

Figure18. Frequency-consequence (latent fatalities) curves for the nuclear chain, based on PSA for the Swiss nuclear power plant Muehleberg (Cazzoli et al., 1993) and on the Chernobyl-specific potential latent fatalities with and without dose cut-off (figure adopted from Hirschberg et al., 1997).

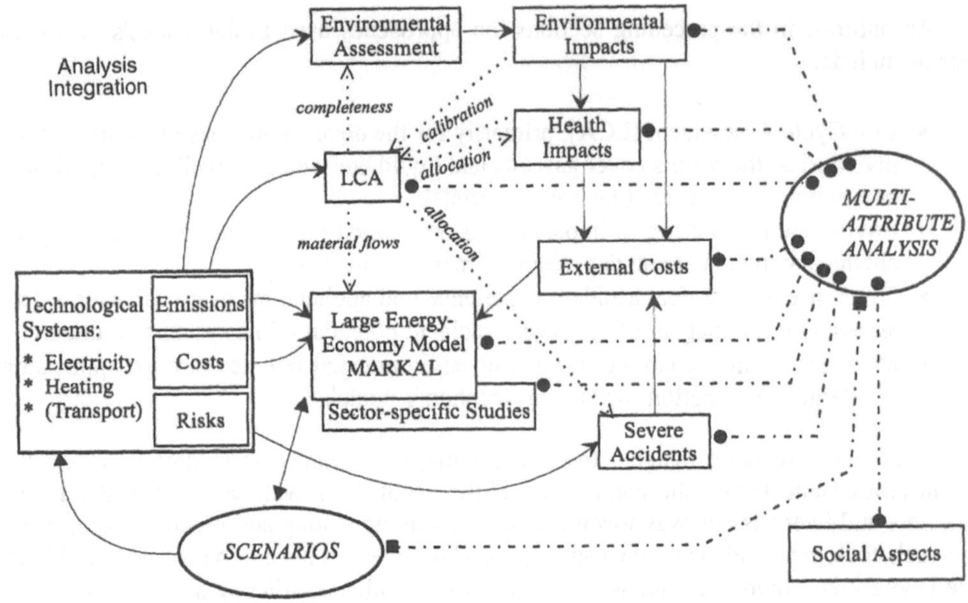

Figure 19. **Principles for analysis integration.**

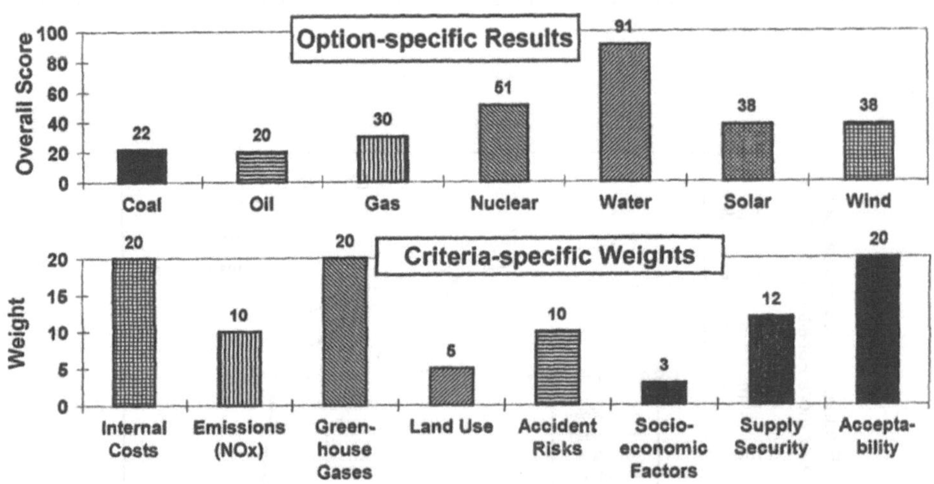

Figure 20. **Reference results of pilot application of multi-criteria analysis to current systems.**

decision-making, based upon a simple weighted multiple attribute value function (Belton and Vickers, 1993). The main role of such a model is to enable decision makers to explore the decision, to learn about their preferences and the way in which they influence the decision.

VISA is ideally suited to problems concerned with the evaluation of few alternatives in the light of information about performance on many criteria. The major steps in developing and using a model of this type for multiple criteria evaluation are:

- Identification of alternatives to be evaluated.
- Identification of relevant criteria and structuring a hierarchy of criteria.
- Evaluation of alternatives with respect to criteria (scoring). For example, the alternative which does best on a particular criterion may be assigned a score of 100 and the one which does least well a score of 0; all other alternatives are given intermediate scores which reflect their performance relative to these two end points.
- Evaluation of the relative importance of criteria (weighting). It is usual to normalize weights so that they sum to 1 or 100.
- Synthesis of the scoring and weighting information in order to arrive at an overall evaluation (ordering of the alternatives). The model is a simple weighted value function, $V_i = \sum w_j x_{ij}$ (summed over all j), where V_i is the overall evaluation of alternative i, w_j is the weight assigned to criterion j, and x_{ij} is the evaluation (score) of alternative i on criterion j.
- Sensitivity analysis - an investigation of the robustness of the decision to changes in scores and weights.
- Decision, (implementation and monitoring).

In practice there is much iteration between the steps listed above.

The pilot application concerns current systems operating under Swiss conditions; however, also the energy chain steps external to Switzerland have been included. The set of criteria is not complete and will be extended in a full scope application. Emissions were used as a surrogate for impacts, with NO_x representing air pollutants other than greenhouse gases (treated separately). Criterion "Accident Risks" refers to objective (technical) component of risk; risk aversion is considered to influence the "acceptability" of various technologies. "Socio-economic Factors" cover employment effects and build-up of knowledge of high value. "Security of Supply" addresses foreign dependence, fuel storage possibilities and achievable load factors. The scores were mostly based on the numerical input generated within the GaBE project. Figure 20 provides the evaluation results for the reference case and the set of criteria used. Weights and scores are shown in the figure.

Numerous sensitivity analyses were performed to investigate the effect of changes in criterion weights and different scores. The sensitivity analyses are extremely important since the results of this type of study should not be viewed as absolute measures but rather as indicators of preference. The ranking of the nuclear option appears to be quite insensitive to variation of all criterion weights with the exception of "Acceptability". This is shown in Figure 21.

The full scale application will ultimately compare scenarios (i.e. technology mixes), with economic potentials of specific systems being considered as a boundary conditions.

CONCLUSIONS

Traditionally, Life Cycle Analysis (LCA) has been applied to operating systems. The present work demonstrates that robust environmental inventories can be generated by a

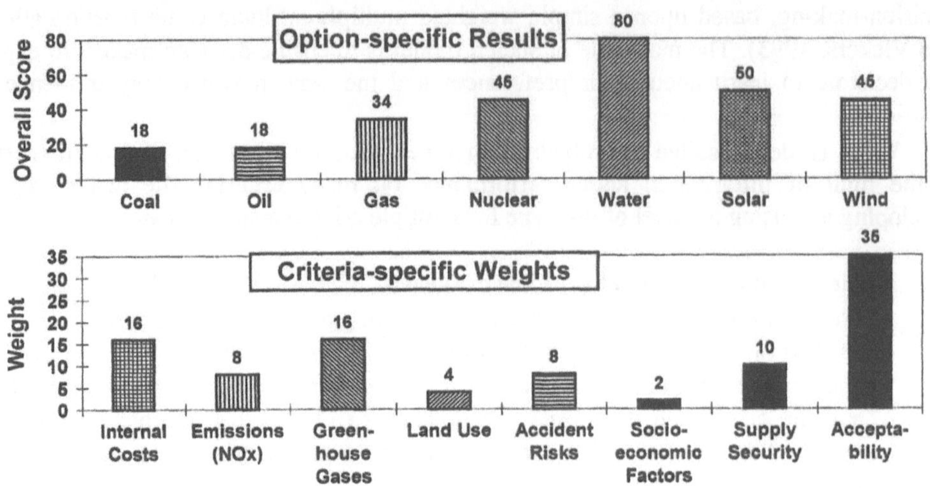

Figure 21. Sensitivity analysis - changes resulting from increasing the weight of criterion "Acceptability".

novel extension of the standard methodology to future systems.

It appears that, in the time span here envisaged, far reaching advances will materialize in fossil power plant technologies. Therefore, the relative importance of other activities in the fossil chains as well as of transportation and material processing increases. As expected, for all systems analyzed the overall emissions of major pollutants are significantly reduced in the absolute sense. As with current systems, the hydropower chain exhibits the smallest emissions of combustion products, followed by nuclear and solar photovoltaic. The improvements in photovoltaic systems in relation to the performance of presently commercially available technologies are expected to be very significant; of particular importance are the reduced electricity requirements in the cell manufacturing processes, leading to the corresponding reductions of "gray" emissions. The radioactive emissions are in relative terms highest for the nuclear fuel chain; however, the relative contribution of the power plant itself is in this context very small.

The results of the systems analyses are clearly reflected in the findings specific for the considered future supply mix options. Thus, supply options which include a significant share of fossil systems (either as a domestic source or as a source of imported electricity) lead to substantial increases in the emissions of combustion products in comparison with the current situation. Nuclear-based options are much more favorable in this context. With respect to emitted radioactivity the nuclear-based options exhibit LCA-results that are one order of magnitude higher than the ones associated with the other options. However, these emissions are lower than the current ones and generally negligible in comparison to the normal background.

As a result of recent efforts the basis for the technical comparison of severe accident risks associated with different energy chains has been significantly improved. This applies in particular to the completeness of historical records, quality and consistency of the information, and coverage of various types of damages.

The evaluation of historical experience demonstrates numerical differences between the aggregated risk indicators obtained for the various energy chains, as well as between the corresponding frequency-consequence curves. Regional differences have been shown to be of utmost importance for the nuclear and hydro chains.

PSA perspective on severe accident risks is particularly important for energy chains where power plants dominate the risks, the historical experience is scarce or its applicability is highly restricted. These conditions are valid for most western hydro and nuclear power plants.

Role of externalities in the context of energy policy and planning is a debated matter although increased attention is currently being given to this issue and to the closely related concept of sustainability. Non-quantifiable aspects that may be of large importance are not accounted for as a part of the estimated external costs. In a decision process, apart from technical and economic aspects, social preferences need to be considered. The actors participating in the decision making process frequently disagree on the objectives and also on assumptions made in the treatment of data and/or in analytical methods. Multi-criteria analysis provides a promising framework that allows the often conflicting socioeconomic and ecological criteria to be addressed simultaneously. Thus, accommodation of "trade-offs" is made possible. Within the GaBE project multi-criteria analysis framework is being implemented. It can be executed by assigning weights to different objective functions (e.g. security of supply, system costs, environmental and health impacts due to normal operation and accidents, etc.), based on advice and recommendations of policy makers, and performing optimal allocation of system resources.

ACKNOWLEDGMENTS

The author is indebted to Mr. R. Dones for his input to the various parts of this paper, to him and Mr. U. Gantner for the work on environmental inventories, to Drs. S. Andreani-Aksoyoglu and J. Keller for their contributions to environmental impacts analysis, and to Dr. G. Spiekerman for his work on the severe accidents database.

REFERENCES

Andreani-Aksoyoglu, S., Graber, W., and Keller, J., 1995, Preliminary simulation of a summer smog situation in Switzerland using the Urban Airshed Model (UAM)", Proceedings of 3rd International Conference "Air Pollution '95, 26-28 September 1994, Porto Carras, Greece, in: Power, H., Moussiopoulos, N., and Brebbia, C.S. (Eds.), Air Pollution III, Vol. 1: Air Pollution Theory and Simulation, Computational Mechanics Publications 343-350.

Andreani-Aksoyoglu, S., and Keller, J., 1996, Influence of meteorology and other input parameters on levels and loads of pollutants relevant to energy systems: A sensitivity study, in: Caussade, B., Power, H., and Brebbia, C.S. (Eds.), Air Pollution IV: Monitoring, Simulation and Control, Computational Mechanics Publications 769-778.

Belton, V. and Vickers, S. P., 1993, VISA, Visual Interactive Sensitivity Analysis for Multiple Criteria Decision Aid, User Manual, Belton V. and SPV Software Products, Edinburgh.

Boone, C. and Howes, H., 1996, Accounting for the environment", CMA Magazine, June 1996.

BUWAL, 1994a, Swiss Federal Office of Environment, Forests and Landscape (BUWAL), Report on the Environment 1993 — The State of the Environment in Switzerland, EDMZ, Bern, Switzerland.

BUWAL, 1994b, Swiss Federal Office of Environment, Forests and Landscape (BUWAL), Globale Erwärmung und die Schweiz: Grundlagen — Umwelt-Materialien Nr.9: Internationales (in German), BUWAL, Bern, Switzerland.

Cazzoli, E. et al., 1993, A Regulatory Evaluation of the Muehleberg Probabilistic Safety Assessment, Volume II: Level 2, ERI/HSK Report 93-304, HSK 11/356 (Limited Distribution), Villigen, Switzerland.

Dones, R., Gantner, U., Hirschberg, S., (PSI), and Doka, G., Knoepfel, I. (ETHZ), 1996, Environmental Inventories for Future Electricity Supply Systems for Switzerland", PSI Report No. 96-07, Paul Scherrer Institut, Wurenlingen and Villigen.

Dones, R., Gantner, U., and Hirschberg, S., 1998, Environmental inventories for future electricity supply systems for Switzerland, International Journal Global Energy Issues, Special Issue on "Knowledge and Decision Support Systems for Integrated Modeling in Energy Policies and Management" (to be published).

DNV, 1992, Worldwide Offshore Accident Databank, Det Norske Veritas Industry AS, DNV Technica, Hovik, Norway.

EC, 1995, ExterneE: Externalities of Energy, Volumes 1 - 5, European Commission, Directorate-General XII, Science Research and Development, Office for Official Publications of the EC, Luxembourg.

Frischknecht R., Hofstetter P., Knoepfel I. (ETHZ-LES), and Dones R., Zollinger E. (PSI) (1994), "Ökoinventare für Energiesysteme — Grundlagen für den ökologischen Vergleich von Energiesystemen und den Einbezug von Energiesystemen in Ökobilanzen für die Schweiz", 1st Ed., ETHZ/PSI, Zurich (1994).

Frischknecht, R., Hofstetter, P., Knoepfel, I. (ETHZ-LES), and Dones, R., Zollinger, E. (PSI), 1996, Ökoinventare für Energiesysteme — Grundlagen für den ökologischen Vergleich von Energiesystemen und den Einbezug von Energiesystemen in Ökobilanzen für die Schweiz, 3rd Ed. (in German), ETHZ/PSI, Zurich, Switzerland.

Glickman, T. and Terry, K., 1994, Using the news to develop a worldwide database of hazardous events, Center for Risk Management, Resources for the Future, Washington DC, USA.

Hirschberg, S., 1995, Framework for and current issues in comprehensive comparative assessment of electricity generating systems, Invited paper presented at the International Symposium on "Electricity, Health and the Environment: Comparative Assessment in Support of Decision Making", organized by EC, ESCAP, IAEA, IBRD, IIASA, OECD/NEA, OPEC, UNIDO, WMO, 16-19 October 1995, Vienna, Austria, IAEA Proceedings Series, Vienna, Austria, 245-278.

Hirschberg, S. and Cazzoli, E., 1994, Contribution of severe accidents to external costs of nuclear power", Proceedings of ENS Topical Meeting on PSA/PRA and Severe Accidents '94, 17-20 April 1994, Ljubljana, Slovenia 242-249.

Hirschberg, S., Spiekerman, G., and Dones, R., 1997, Comparative assessment of severe accident risks in the energy sector", The 1997 Annual Meeting of the Society for Risk Analysis - Europe, Stockholm, Sweden, 15-18 June 1997.

Hirschberg S., Spiekerman G., and Dones R., 1998, Severe Accidents in the Energy Sector (to be published as PSI Report in 1998).

Houghton, J.T. et al., Eds., 1995, Climate Change 1994 — Radiative Forcing of Climate Change and An Evaluation of the IPCC 1992 IS92 Emission Scenarios, Cambridge University Press, Cambridge, UK.

IAEA, 1992, Comparative Assessment of the Health and Environment Impacts of Various Energy Systems from Severe Accidents, Working Material, Proceedings of a Technical Committee Meeting Organized by the International Atomic Energy Agency and held in Vienna, Austria, 1 - 3 June 1992.

IEA, 1993, Energy Statistics and Balances, OECD (1960 - 1991), Non-OECD (1971 - 1991), International Energy Agency, Diskette Service, OECD, Paris, France.

Kroeger, W., 1998, Sustainable development of energy supply", in this Proceedings.

Lee, R., Hirschberg, S., Boone, C., and Dutkiewicz, R., 1997, General Guidelines for Comparative Assessment of the Health and Environmental Impacts of Electrical Energy Systems" (to be published by the IAEA in 1998).

Moosman, K. L., Goldman, M., Massé, F., Mills, W. A., Schiager, K. J. and Vetter, 1996, Health Physics Society Position Statement: Radiation in Perspective, HPS Newsletter XXIV.

ORNL&RfF, 1992 - 1995, U.S. - EC Fuel Cycle Study, Reports 1 - 8, prepared by Oak Ridge National Laboratory and Resources for the Future, McGraw-Hill/Utility Data Institute, Washington DC, USA.

Rowe, W., 1977, An Anatomy of Risk, John Wiley&Sons, New York.

Swiss Re, 1994, Sigma, Swiss Reinsurance Company, No. 1/2, January/February 1994.

VSE, 1995, Verband Schweizerischer Elektrizitätswerke (VSE), Vorschau 1995 auf die Elektrizitätsversorgung der Schweiz bis zum Jahr 2030" (in German and French), VSE, Zurich, Switzerland.

LEGISLATIVE FRAMEWORK NECESSARY FOR THE SUCCESSFUL REEMERGENCE OF THE NUCLEAR INDUSTRY

Michael T. Woo
United States Enrichment Corporation
Finance Department
6903 Rockledge Drive
Bethesda, MD 20817

INTRODUCTION

Perhaps it is because of my background in energy and nuclear economic analysis in Congress that I have been invited to speak about the legislative framework necessary for the successful reemergence of the nuclear industry. So I will act like the economist who is asked how he would suggest hanging a picture. As any good economist, he first states his assumptions: "First, assume the existence of a hammer and a nail...". In this case, I will assume the existence of a nuclear technology that is environmentally benign and economic. I will leave to you to ponder whether this would be a new or existing technology. I am not going to get into the "assumptions". Rather, I am going to focus on how to hang the picture if one has a hammer and nail -- what is the legislative framework which would allow this nuclear technology to succeed.

PARADIGM EVOLUTION: OUT WITH THE OLD AND IN WITH THE NEW

Now, let us consider that there are two basic paradigms to consider for the legislative framework for nuclear power. One is the traditional and historically applied "command and control" paradigm. The new paradigm is market determination. Slowly but surely the legislative framework is moving away from the old paradigm toward the new paradigm of reliance upon market determination.

The command and control paradigm has the nuclear industry regulated stem to stern, the rate of return is rate base regulated top to bottom. Safety regulation is micro-managed. Environmental solutions are mandated.

The success of the command and control paradigm has been subject to dispute. The industry argues that the plants have operated safely with little or no release of radioactivity. On the economic front, they can show that most plants are economic, particularly if you considered the marginal cost of operation. Environmentalists have argued that the waste problem has not been solved; that the safety of the plants is suspect and that on a full cost basis, most existing nuclear plants are not economically competitive at the current market price. The emergence of deregulation as a driving force has shifted the emphasis away from command and control, toward market forces.

The market determination paradigm has the technology privately financed -- not rate based financed. Returns are unregulated and not guaranteed. Safety and environmental regulation are result based.

Some may argue it is hard to see how the nuclear industry could survive under such a legislative framework. Those skeptics must consider the "economist assumption". I am assuming that the nuclear technology is economical and even more environmentally benign that the current generation of power plants. The market forces act like a sieve. In other words, the market framework would allow such a technology to succeed. It is also true that a market-based legislative framework operating like a sieve would not allow the nuclear industry to succeed if the assumption of low cost environmental friendliness were not true.

The desired legislative framework depends on how far toward the market paradigm it is desirable to go. I will describe effects of this paradigm change on legislation and the industry.

ECONOMIC FRAMEWORK

Existing nuclear plants in the U.S. have emerged under the well known legislative framework of state-based, vertically integrated, rate based economic regulation. Historically, large nuclear power plants thrived under this form of regulation. The high initial capital cost and low operating cost combined for a strong economic incentive to build nuclear. The utility industry argues -- and I believe correctly -- that they were also encouraged to build nuclear plants by their regulators and the government for fuel diversity, base-load dependability, and environmental factors.

In the late 1970's and early 1980's, the legislative framework for economic regulation changed. With the passage of the National Energy Act in 1978 and the Presidency of Ronald Reagan, the economic framework slowly shifted to market forces. This shift was continued in The Energy Policy Act of 1992 and further continued with electric restructuring action in states all around the Nation. During the consideration of the Energy Policy Act of 1992, in my prior role as a Congressional staff member, I had numerous discussions with those in the environmental community who were insistent that there were economic alternatives to nuclear power. In the end, we agreed that a more market-oriented framework would allow the more economic energy technology to succeed and penalize the less economic energy technology. Thus, in the end, the paradigm shift was supported both by industry and environmentalists.

The revised legislative framework is moving to allow energy technologies to compete head-to-head in the market for power. At the state regulatory level, the recovery of so-called stranded costs, which allows existing nuclear plants to compete on the basis of their lower variable cost, is controversial work-in-progress. The result is that some nuclear plants will close before license renewal. This case by case, state by state basis is creating an uneven playing field for existing nuclear power plants.

As a result, if my basic "economist" assumption is correct, i.e., the existence of a environmentally benign, economic nuclear technology, the market based legislative framework would permit new nuclear plants who meet the test to succeed. However, the reverse is also true. The market-based framework would prevent an uneconomic technology from succeeding. If you assume a successful technology, the market will allow it to succeed. Pretty smart those economists!

SAFETY

As stated previously, the legislative framework for safety regulation of the nuclear industry has remained detail oriented as opposed to results oriented. As the U.S. moves to a more competitive environment where innovation is a key to success, flexibility will be essential for any energy technology to succeed. One step licensing for the new generation of advanced reactors is a move in this direction, yet it is unclear whether any technology can succeed if it is saddled with burdensome regulation of minute details.

It may be that the severe adverse consequences resulting from the risk of failure requires this micro-approach in regulation. The cost of catastrophic failure of each step of the process or each component may be too great to allow for result regulation even at the detail level. If so, then my first assumption of an economic technology will require new nuclear plants to be economic even with micro-managed safety regulation.

The industry consistently emphasizes the point that it isn't an absolute from a safety point of view that micro-management is a priori the best way to proceed. Detailed micro-management may limit the ability to see the big picture or to apply "common sense" reactions to problems, or to permit needed flexibility. An oft cited example is requiring a plant to be shut down for minor changes or repairs instead of permitting them to be made while the plant continues to operate. Similarly, it puts great burdens on the regulator to know exactly what they are doing. They need to keep track of what is going on both at the detail and big picture levels. The other effect of micro-management is people-intensive; the regulator needs more and more technical staff and expertise. It is hard to believe such regulation will be sustainable outside the U.S. and a few well-developed nations.

If the framework of safety regulation is to change, it will require legislation. It required legislation for one step licensing to proceed. Any such change would be a major change in safety implementation and would require consensus. The industry should realize that a strong nuclear safety regulator -- both in reputation and deed -- is important for the success of the industry. However, a strong regulator can ensure safety in a way which is cost effective.

One interesting market-based approach the environmentalists have proposed is the elimination of Price-Andersen coverage for new nuclear plants. This would force internalization of risk to the new industry. Could this be exchanged for result-based safety regulation in a legislative framework? Would the industry accept such risk? If the technology is economic, it should be able to absorb such a risk. But it may not be able to absorb such a risk if it is still micro-managed.

ENVIRONMENT

Why am I talking about the environmental framework when I assumed the technology was environmentally benign? Well, environmental legislation can affect whether even an environmentally benign technology can succeed. The major environmental issue facing nuclear technology currently is lack of action on interim and final disposal of nuclear power plant wastes. In some states, there is an explicit bar to new nuclear plants unless the waste disposal issue is resolved.

Clearly, a legislative framework is needed for this environmental problem to be resolved. And today we are at a critical juncture in that process as Congress votes on making Yucca Mountain an interim storage site to receive the Nation's spent fuel. While it can be argued that existing plants can store their waste, it is unlikely that any new plants will be built with the prospect that waste will reside at the facility for the foreseeable future.

However, here too, the legislative framework can legislate the details or legislate the results. It is not obvious that the government can do waste storage any better or worse than the private sector. The use of market forces within a regulatory framework may more successfully solve this problem -- or make the costs of solving the problem more evident.

INTERNATIONAL

While the U.S. has the luxury of considering alternative paradigms for legislation, many of the nations currently considering nuclear will have a more limited choice. NRC type of regulation assumes an NRC. Perhaps, this is not possible in China, India, Brazil or North Korea. For the nuclear industry to reemerge on the world stage, the statutory framework for nuclear power must be considered in each nation.

Given the lack of a regulatory infrastructure comparable to the NRC in scope, a more market oriented statutory framework is likely to be adopted. If my assumption of a new, environmentally benign, economical technology is correct, this framework will support the reemergence of the nuclear industry in the world energy market. However, it is also the case that a market oriented framework would also ensure failure if it is not correct.

CONCLUSION

In order for nuclear power to succeed in the United States, a consistent legislative framework is needed. That framework could be highly regulated or market based. In my view, it is more likely to be market based. However, in order for nuclear power to succeed it is also true that economists' assumption of a hammer and nail must also be correct -- there must be an environmentally benign, economic nuclear technology. Unless new nuclear plants are both economic and environmentally benign, they can not succeed under a market based legislative framework.

The final consideration is public acceptance. Any legislative framework is not possible without public acceptance. There is something of a "chicken and egg" problem here. Unless Congress acts to establish the appropriate legislative framework, the nuclear industry cannot move forward and improve the public acceptance. Perhaps through forums like this, a consensus can be formed on a legislative framework that is worthy of public acceptance and that the industry would rise or fall on the basis of its success of operating within that framework.

I look forward to your thoughts on the appropriate legislation framework and how we can go about achieving it.

SECTION IV Energy Future of the World

THE PREDICTED DEMISE OF U.S. NUCLEAR POWER CAN WE SAVE IT FOR OUR FUTURE WELFARE?

Bertram Wolfe

For the past twenty years nuclear energy plants under construction in the U.S. have encountered such regulatory and legal delays that they have taken a dozen to twenty years to complete, with resulting uneconomic costs. As a result not a single Electric Utility intends to build a new nuclear plant in the U.S. . Indeed, a few years ago when the NRC requested aid just to demonstrate the feasibility of a new nuclear plant siting procedure, no Utility was willing to help - even though it was made clear that there was no commitment to actually build a new plant.

Overseas, in Europe and Asia the situation is different. It takes only 4 to 6 years to build a plant of U.S. design abroad. Last month, in Japan, a ceremony took place celebrating the commercial operation of two new U.S. designed nuclear plants which were each built in four years, and operate economically. Indeed, in order to reduce fossil fuel burning so as to help alleviate the potential Earth Warming problem, Japan is considering building 20 additional nuclear plants in the next dozen years (on top of their present 50), and there are similar plans in the growing economies around the world.

The U.S. nuclear construction delays have resulted from an inefficient, bureaucratic licensing system; and a court system which has imposed long delays in response to anti-nuclear legal challenges. These continuing problems were recognized by the Nuclear Regulatory Commission (NRC), the agency which is responsible for nuclear plant safety and licensing. In 1986 NRC Director Landau Zech initiated development of a new "Standardized" licensing procedure. This requires that the design and construction plans for a new nuclear reactor be reviewed in fine detail, with public comments considered, before a license is issued. In principal then, once the license is issued the plant construction can proceed with no delays as long as it adheres to the licensed design. And this year "Standardized" plant licenses were finally issued to General Electric (GE) and Combustion Engineering (CE); it is anticipated that Westinghouse (W) will receive a license next year.

Dr. Bertram Wolfe is a forty year veteran of the peaceful nuclear energy field where he has been a leader in technical areas and in the design of a number of nuclear reactors. He is a member of the National Academy of Engineering, a past president and fellow of the American Nuclear Society, a Professional Engineer in California, and has received a number of industry awards and honors. He is the author of over a hundred papers on energy and nuclear energy. Dr. Wolfe retired from General Electric in 1992, where he was a Vice President and head of GE's peaceful nuclear energy business. He has since been an independent consultant, and has served on several Board's of Directors and a number of academic and industry advisory boards

Environment and Nuclear Energy
Edited by Kursunoglu *et al.*, Plenum Press, New York, 1998

One might note that GE's new licensed nuclear plant design is the same as that for the plants recently constructed in Japan in 4 years, and that CE's plant design is similar to the CE plant designs built in Korea in 5 years. But who would risk over a billion dollars on the assumption that U.S. designed nuclear plants can now be built here as efficiently as they are abroad? One can be sure that anti-nuclear groups will file a number of law suits when a new nuclear plant is initiated. Who can be sure that onerous, costly, legal delays will not occur, despite the intent of the new licensing procedure?

The Energy Information Agency answers these questions by projecting that in the next 20 years over half the nuclear plants now operating in the U.S. will be shut down as their operating licenses expire, and not a single new nuclear plant will be built to replace them.

Since, except for hydroelectric dams, nuclear power is the only major electricity source that does not produce Greenhouse gases, the net effect will be significantly increased U.S. Greenhouse gas generation; this at a time when the world is concerned about environmental disasters from the warming of the earth.

In addition, if the nuclear industry in the U.S. goes downhill as projected, the nation faces loss of its international leadership and influence in assuring safe nuclear reactors world wide (no reactor in the world built to U.S. standards has harmed a single member of the public); and loss of its ability to help assure that nuclear materials will not be diverted from nuclear power reactors to military uses.

As indicated by experience abroad, U.S. nuclear problems are institutional, not technical, and are present at a time when the welfare of the nation and the world may depend upon a major expansion of nuclear power. To maintain our world leadership, and be able to maintain and expand our nuclear energy capability here, the U.S. government should *expeditiously* demonstrate that it has corrected its licensing problems by supporting the early construction of at least two new nuclear energy plants. At the least it should provide monetary insurance against government and court construction delays, while hopefully demonstrating that such delays will not occur, and that U.S. plants can now be built here as efficiently and economically as they are abroad.

In the coming half century, even with a major successful conservation program, world energy use is projected to triple, due to the doubling of the 4 billion population of the third world and their increasing standard of living. *Without a major world expansion of nuclear power we face potential international conflicts over scarce energy supplies, and the potentially disastrous warming of the earth.* Maybe fusion or cold fusion, or solar or wind power, or some new major energy source will be developed. We should work on them. But their problems of large scale practicality are so great that it would be irresponsible to count on them.

Thus, this nation should assure that we maintain our nuclear energy capability and leadership. They may be vital to our future welfare and the welfare of our children.

ENVIRONMENT, NUCLEAR ENERGY, AND PUBLIC PERCEPTION

Hans Blix

Director General Emeritus
International Atomic Energy Agency
Vienna, Austria

The famous Indian nuclear scientist, Dr. Homi Bhabha, coined the expression "no energy is more expensive than no energy". He meant that nothing is grimmer than living without any other energy than that of your own body. For a good life we need energy for cooking, heating, transportation, industrial production, communication, home appliances ...

We take our cars, elevators, dish-washers, central heating and computers for granted in the industrialized societies. When we need power, we just plug in. Our energy and especially electricity consumption is big. The US uses some 13,500 kWh per person and year; my own country, Sweden, some 15 000. But at the other end of the spectrum we have countries like Chad and Tanzania with less than 100 kWh per person and year. Nothing is more certain - and reasonable - than that countries at the lower end will seek to improve the lives of their people by increasing their energy and electricity consumption.

China - with over a billion inhabitants - is planning to increase her electricity generating capacity by around 16 GW/year - or sixteen 1000 MW power plants. Most of this capacity may be coal-fired. Presently China has three nuclear power plants in operation and six more in the pipeline. India is planning to increase her electricity generation by some 200 GW by the year 2020 - i.e. 200 large power plants - again, it may be mostly coal fired. Considering the substantial increase in CO_2 emissions such electricity expansion programmes would cause, we would have reasons to wish good progress in the nuclear programmes of these countries. Solar cells or windmills are not likely to be alternatives to the coal-fired plants.

If the wish for a better life is one factor driving increased energy demand, population increase is the other. It has been calculated that at the time of Christ there were some 350 million people on the earth; by the year 1900 there were 1.5 billion; in 1990 we were 5 billion and we expect to be 6 billion by the year 2000. Thus in the last ten years of this century we are increasing almost as much as we did during the first 1900 years after Christ. Although there has been a significant slow down in the last decades, one must ask whether this is a "sustainable development" pattern. Where are we moving? To "standing room" only?

At this juncture, when fast developing countries are getting into gear to increase their energy use, we are discovering that, unfortunately, the largely fossil-fuelled energy fiesta which we have enjoyed so much in the industrialized world cannot continue. While the Rio Conference on Environment and Development in 1992 recognized that energy is essential to economic development, it declared that "the current global pattern of energy supply and use is not sustainable. *Environmentally sound energy approaches* are necessary to control atmospheric emissions of greenhouse and other gases and substances". Which are these "environmentally sound energy

Environment and Nuclear Energy
Edited by Kursunoglu *et al.*, Plenum Press, New York, 1998

approaches"? The Rio Conference did not say, except that it expressly endorsed energy saving and modern renewable sources, like solar and wind power and biomass. Since Rio the dangers of our present global pattern of energy use seem to be on ever firmer scientific ground, but agreement on responses does not seem to be any closer.

I do remember that Groucho Marx once said: "Why should we do something for future generations? They never did anything for us ...". Well, he meant to be funny - as indeed he was. At any rate we do not want to damage life conditions for future generations.

There is no great mystery about the environmental threats coming from our increased energy use. While large hydro dams often cause opposition on the ground that they take away large areas of land from local populations, the major concerns are with the expanding use of coal, oil and gas. Thirty years ago the concerns related to acid rains, dying forests and lakes due to emissions of SO_2 and NO_x. These emissions, as we know, *can* be removed or reduced - at a price - but in large areas of the world are not.

Today's chief concern is related to CO_2 emissions which along with other so-called "greenhouse gases", such as methane leaking from gas fields and pipelines, are believed to contribute to "global warming". There does not appear to be any viable technique in sight to remove or neutralize the CO_2 which is formed in the burning of all fossil fuels.

It is curious that in this situation we do not have much of a discussion between governments about energy policies. To be sure, what energy mix we rely on is decided inside States, not between them. And the mix varies considerably. In Norway, which uses more electricity per person than any other country in the world, practically all the electric power, so far, has come from abundant hydro power. In Poland, as in Australia, rich in coal, most electricity is generated by the burning of that coal. In my country, Sweden, about half the electricity has been coming from hydro and the other half from nuclear power. Etc.

Of course the choice of national energy mix has an effect on the aggregate global energy mix and the global emissions of greenhouse gases, in particular CO_2. But at present these national energy mixes are left out of the intergovernmental discussions. Instead governments have chosen to discuss *targets* for restraint in greenhouse gas emissions, leaving it to themselves to decide on energy and tax policies which could lead to fulfilment of the targets. As often as these targets have been proclaimed, they seem to have had little effect:

- In 1988, the so-called *Toronto target* wanted to "reduce CO_2 emissions by approximately 20% by the year 2005". However, since that target was adopted in 1988, the global CO_2 emissions have, in fact, increased by some 16%;
- In 1992 there was the Rio target for industrialized countries to return to 1990 levels of CO_2 emissions by 2000. However, since that target was set, CO_2 emissions in OECD countries have, instead, increased by some 8%;
- This year, at the special session of the UN General Assembly, a 15% reduction of greenhouse gases by 2010 compared to the year 1990 was urged. There is at present much discussion about this target and somewhat more modest targets proposed by Japan and the US. However, it should not go unnoticed that a study by the International Energy Agency (OECD) projects global CO_2 emissions in 2010 to be 36 to 50% above their 1990 levels, with OECD emissions also higher than in 1990.

It is against this background that the Kyoto Conference on the Climate Convention will take place in a few weeks time and that G-8 Energy Ministers will meet in Moscow next year. As if the difficulty were not great enough to consider greenhouse gas restraints in the OECD group, there is now a growing risk of a North/South confrontation.

Simplified a bit, the arguments could go a little like this. Industrialized countries might say to the developing countries:

"We are sorry - without knowing the risks we have already put far too much greenhouse gas into the earth's atmosphere in our quest for decent standards of living. We will try to restrain ourselves a bit - especially by energy efficiency and taking advantage of the available gas, which releases half as much CO_2 per energy unit as does coal. But for heavens sake, now that we all begin to understand how dangerous these excessive emissions may be, don't you race ahead with a vastly increased burning of fossil fuels, especially coal. We only have one atmosphere. If you race ahead, you put us all in jeopardy - including yourselves. Indeed, we might not make any commitments at all unless you do it, too ..."

Developing countries, especially the fast developing populous ones, in Asia might reply:

"Are not all human beings equal? Do we not have the right to raise the standard of living for all our people by using more energy as you have done? If the earth's atmosphere has a limited capacity to absorb greenhouse gases without risk to the global climate, let us share it equally - equal emissions per person!"

It will certainly not be easy to reach some meaningful agreement. Meanwhile the already excessive greenhouse gas emissions increase further. It seems to me that *at some point* this dilemma must lead to serious discussions at the governmental level about energy *sources* and their relations to greenhouse gas emissions. And perhaps to a greater awareness that nuclear power could provide increasing amounts of base load electricity *without* adding greenhouse gases. At the General Conference of the IAEA in October some government representatives did, indeed, acknowledge this. To take some examples, the representative of Japan said (29 Sept. 1997):

"... In our view, nuclear power will play an important role in responding to the question of global warming. Provided that safety is ensured, we look to nuclear power as a realistic energy option as it excels in supply stability and offers low environmental impact free from greenhouse gas emissions".

The representative of the United States, Secretary of Energy Federico Peña, said (29 Sept. 1997):

"It is essential that we remain capable of ensuring the safety of our nuclear reactors. With populations and standards of living increasing around the globe, nuclear energy could play a potentially significant role - helping the world meet an ever increasing demand for energy while also helping to reduce emissions of greenhouse gases".

The representative of Canada said (1 Oct. 1997):

"Nuclear energy is a safe, environmentally sound and cost-effective source of energy. Canada is a firm supporter of the nuclear energy option which is an important component of a sustainable energy supply mix for many countries. Among its many advantages, nuclear power significantly reduces emission of greenhouse gases and other noxious gases that otherwise would have been emitted to the detriment of the environment and or human health".

These statements are supported by solid data. We have calculated in the IAEA that if the present some 440 nuclear power reactors in the world were closed and replaced by fossil-fuelled plants, there would be an *8% increase* in global CO_2 emissions from energy.

I could also cite the Executive Director of the International Energy Agency of the OECD, Mr. Priddle, who noted that "nuclear power accounted for the greatest part of the lowering of carbon intensity of the energy economies of the OECD countries over the last 25 years".

Such data and comments were conspicuously absent at the special session of the UN General Assembly on sustainable development in June. As far as I have been able to see, I was, myself, the only speaker to suggest that the nuclear power option could be important to help restrain greenhouse gases. Government speakers were silent on the matter. A representative of Greenpeace did, however, pronounce himself. He called for, I quote:

"the beginning of the phase out of fossil fuels and their replacement by clean renewable energy which exists in abundance". He added, "nuclear is not an alternative ...".

Let us remember that fossil fuels constitute 85% of the world's commercial energy, nuclear power about 6% and renewables less than 2% - most of it geothermal. Yet, many of the governments, like Greenpeace, seem to suggest in the UN - but not in the IAEA - that salvation lies in the non-hydro commercial renewables and in energy efficiency. With what conviction, I wonder ...

Let me first discuss the hopes pinned on solar and wind power and biomass. There is a good deal of support for these sources by governments around the world and there would certainly be a lot of money to make in technical breakthroughs making these sources economically competitive. It is surely not lack of resources for research and development that is the reason why these energy sources do not constitute even 2% of the world's energy. It seems more likely that their low energy density makes it hard for them to become competitive with fossil fuels or nuclear power for electricity production.

It is not suggested that it would be technically impossible to develop huge wind farms, vast biomass plantations or fields of solar cells. Some both electric and non-electric niche applications have been found. For instance, solar energy is put to good use for the heating of household water and photovoltaic cells are used for electricity generation in distant places where it would be difficult or costly to build power lines. However, it is hard to imagine that these sources of energy could become economically viable for large-scale base load electricity generation. The reason is that harvesting energy sources that have low density is expensive. Achieving the electricity generating capacity of a 1000 MW(e) power plant by these sources requires cumbersome installations and operations:

- an area in the range of 25 to 50 km^2 of solar cells;
- 50 to 150 km^2 for windmills; or
- 3000-5000 km^2 of biomass plantations.

In spite of the progress made, investments remain high for solar cells and windmills and the intermittent character of solar and wind energy is a severe handicap so long as we do not have better means of storing electricity. It is hard to imagine the megacities - like Mexico City, Beijing, Jakarta or Calcutta - relying on intermittent and low energy density renewables.

Increasing energy efficiency - and thereby restraining CO_2 emissions - is another matter. It can be done from the stage of fuel extraction, through energy generation and to the end use. No one is opposed to it, but it does not happen overnight, but in the context of new investments. It will not, however, by any means neutralize the increasing demand for energy. The Chinese now use about 1000 kWh per person and year. We cannot expect them to try to attain Japanese standards of electricity services simply by increasing efficiency.

I am convinced that in the desperate search for ways of expanding the global energy supply without adding to the risk of global warming, governments will sooner or later rediscover nuclear power. I recall a story I once heard about a group of scientists on a boat somewhere in the Pacific. They had lost their navigational tools and in desperation they asked an islander helping hand on board in what direction he thought Bora Bora lay. The man unhesitatingly pointed in one direction and after 30 hours steaming on the basis of this advice they sighted Bora Bora. They asked him how he knew that Bora Bora was there. He answered: "Bora Bora always was there ..." Those of us who have some familiarity with the nuclear island are also tempted to tell fellow passengers who seem at a loss to know in what direction they should go, that the nuclear island is there within reach all the time ...

When will this advice be listened to? I don't know. I think governments must first recognize that they really stand almost empty handed in the face of the risk of global warming. Governments cannot, of course, come out overnight as advocates of nuclear power. They - and we - must begin by creating a more accurate understanding and image of the benefits and risks of nuclear power. However, time is of essence. The energy choices we make today decide the composition of the atmosphere for a long time, because the plants built today may have lifespan of 40-50 years. Thus it matters if coal plants decommissioned today are replaced by new coal plants or by gas plants or nuclear plants. Conversely, it matters if nuclear plants decommissioned today are replaced by new nuclear plants or by coal or gas plants. And it does matter what new power capacity China, India, Indonesia and other populous countries choose.

Even though the environmental aspects of the energy options today are central, I am not suggesting that the case for the nuclear power option should rest only on the fact that it hardly emits any CO_2, SO_2 or NO_x.

The nuclear option must also remain economically attractive. While this may be easy vis-vis non-hydro renewables for base load electricity, it will constitute a challenge as regards power supply where piped gas is available.

As after the oil crises the energy independence that can be provided by nuclear power for energy poor industrialized countries like Japan, France, Sweden or Korea, remains a great asset. Nuclear fuel for several years of operation can be stored without great difficulty. By contrast, you do not have energy independence at the end of a gas pipeline from a foreign region. Armenia had only two hours of electricity per day when the gas pipeline from Azerbaijan was cut. Only the re-opening of one of the nuclear reactors at Medzamor restored the lights.

When nuclear power is written off, as it has often been in environment dominated UN discussions, it has usually been by reference to "public concerns" relating to safety, waste and non-proliferation. I find it puzzling and unacceptable that a major energy option can be left out simply by a reference to a public opinion without substantial discussion of whether there are *justifications* for this opinion.

Apart from that objection, what about the nuclear safety, the nuclear waste and non-proliferation?

No energy is to be had at zero risk and I am not suggesting that nuclear power is without some risk. However, that risk must be compared with the risk we would take if choosing another source of energy. The largest numbers of casualties from energy related accidents have been linked to the bursting of hydro dams ...

The fall-out from Chernobyl - the radioactive and especially the psychological - demonstrated that "a nuclear accident anywhere is an accident everywhere". This globalization is now meeting a globalized response: all nuclear operators now work together in the World Association of Nuclear Operators to learn from each other and to establish and uphold high safety standards. And within the IAEA an international nuclear safety culture, consisting of binding basic safety rules and more detailed recommended standards and of extensive expert services, is promoted. Under the latest conventions the State parties will also have the opportunity to insist on the full implementation of these rules through mutual peer review concerning their respective nuclear activities.

The increasing safety and reliability of nuclear technology can be seen in the improved production figures for nuclear power around the world, lower doses to plant personnel and fewer unplanned outages. The availability factor for the world's nuclear power reactors is now close to 80%. Unplanned outages, on average below 5%, compares favourably with that of fossil fuelled plants.

The issue of nuclear waste is one in which there is a great distance between public perceptions and reality. There is an international consensus - now embodied in a new convention on basic principles for the safe disposal of nuclear waste and the safe management of spent fuel. The nuclear community knows very well what to do. The serious problems facing us in those places are political - not technical. Nuclear energy is following a path of *confinement* for its waste disposal, with hazard decreasing with time, while the waste from fossil fuels is dispersed into the atmosphere, and placed on the surface of the earth, with some pollutants remaining toxic forever. We know that it is difficult in all countries to find acceptance of sites for nuclear waste disposal. However, when it comes to the waste from coal, oil and gas, nobody asks anybody's acceptance. The site is the atmosphere we breathe or the earth we walk.

Finally, there is no doubt that the acceptance of civilian nuclear power has suffered from people associating to the military use. It has not helped to point out that there can be nuclear power without bombs and bombs without nuclear power. China had the bomb long before it had a power reactor. And many countries have power reactors and no bombs.

In the wake of the global detente, fortunately, there is growing nuclear disarmament and the peaceful atom is gradually more present in our minds than the shrinking stocks of nuclear weapons. The IAEA is now discussing with the US and Russia how its inspectors can verify that fissile material - Pu and HEU - from dismantled weapons is safely stored or peacefully used. The talk is no longer about the MAD - mutually assured destruction. Rather about the more attractive perspective of making electricity out of the bombs. We still have a considerable way to go, however, to minimize the arsenals of the nuclear-weapon States, to persuade some States to roll back from the nuclear weapons - as South Africa did - and to prevent any further proliferation. However, I believe the President of Brazil was right when he recently expressed the view that the era of nuclear weapons is drawing to a close.

Competition between great powers and groups of nations will continue but it will probably be pursued in the fields of economy and information - and efficient, economic and environmentally acceptable energy production. In this perspective I am optimistic about nuclear power. As someone said: governments will act rationally - once they have exhausted all other options.

OVERREGULLATION AND OTHER PROBLEMS OF NUCLEAR POWER

Richard Wilson
Mallinckrodt Professor of Physics
Harvard University

As the world awaits the discussions and agreements at Kyoto, one fact stands out - that of all the alternate fuels nuclear power is alone in producing no appreciable particulate air pollution, not contributing global warming and, with a breeder reactor being able to produce power for 100,000 years at modest cost. The present problem is that the cost has risen considerably in the last 25 years and begins to approach the costs of some of the solar energy alternatives.

25 years ago, the Maine Yankee nuclear power plant had just been completed for $180 million (CMP 1972), or $200 per kwe installed capacity. Connecticut Yankee nuclear power plant was producing electricity at 0.55 cents per kwh busbar cost (Webster 1972), some part of which was paying off the mortgage. Now, 25 years later, the most recently completed nuclear power plants cost at least $2000 per installed kwe of capacity, 10 times the 1972 cost, and Connecticut Yankee is being permanently shut down because it costs 3.7 c/kwh, 6.7 times the 1972 cost even though the mortgage is fully paid (NYT 1997). Yet inflation can only account for a part - perhaps a factor of 2.5 to 3 - of this.

In most technologies there is a learning curve and things get cheaper as time goes on. Roughly speaking one expects a 30% reduction in cost for a doubling of capacity. In this technology we have an unlearning or forgetting curve. Numbers that I have seen from France give an average cost of nuclear electricity including all costs, of 2.9 cents/kwh (Lochak 1995), whereas a similar number in the US averaged over all plants operating in 1995 was 5 cents/kwh (this ignores costs of plants, like Shoreham, which were abandoned for political or other reasons). Why is it now more expensive in the USA? What has happened? We should demand detailed study of this question.

Various ideas include the following:

- in 1970 manufacturers built turnkey plants or otherwise sold cheap reactors as loss leaders. But this can only account for a small proportion of the capital cost.

- Construction costs generally have risen in this time.

- it may be that in 1972 we had good management and good technical people. But why has management got worse when that has not been true for other technologies?

- it is probable that nuclear power plants are safer today than they were in 1972. But it would be hard to argue that the actual safety improvements have cost that much money. Most are a result of more careful thought using such approaches as event tree analysis.

Environment and Nuclear Energy
Edited by Kursunoglu *et al.*, Plenum Press, New York, 1998

- A typical nuclear power plant employs 1300 people today compared to the 250 each in Dresden II and Dresden III many years ago (Benkhe 1997). Is this necessary? Is safety or security really improved?

- many people have suggested that the problem is the over regulation by the Nuclear Regulatory Commission which has no mandate to keep power plants in operation - only to make them safe. That for example is an implication of the Towers Perrin report of 2 years ago (Towers 1995). Why should this be so? I suggest some problems below.

- the lower total costs for the French reactors assume a rate of return on capital of 5% compared to a typical US rate of 10%. Yet the high cost of electricity from Connecticut Yankee cannot include interest on original construction any longer.

- a part of the cost just after the Three Mile Island accident may have been due to large capital retrofits demanded by NRC. No one has suggested such a large retrofit for the fleet of Boeing 707s although it is clear that the latest 757 planes are far safer, and no one has suggested that every coal plant be retrofitted with a nuclear boiler that emits no particulates and not merely a few.

Any of these suggest that it is possible to return to a lower cost if we have the will. What does that involve?

The first Commission of the NRC set some radiation and safety guidelines in the rule making in RM-30-2 in 1974 which followed two years of public hearings begun by the former Atomic Energy Commission. The proposed that expenditure on radiation exposure reduction should be made if it costs less than $1,000 per ManRem. It is noteworthy that this is a larger number than suggested by ANY of the environmental groups participating in the hearings. In 1994 this figure was doubled to $2,000 (Kress 1994). But an independent study (Tengs et al 1995) suggests that NRC regulations for RADWASTE have resulted in expenditures of 1000 times this amount. This is a clear demonstration that the regulations are stricter than necessary and it is likely that this is a reason for an increase of cost.

The Advisory Committee on Reactor Safeguards (ACRS) has repeatedly stated that it is not sensible to regulate on the basis of a PRA (Probabilistic Risk Assessment). But a PRA can be used to discuss whether reactors that were designed and operate under existing regulations are safe enough. But a use of PRA in this way suggests that NRC must act to modify its regulations and make them more or less stringent in accordance with retrospective study. According to this view of the uses of a risk analysis, the NRC RADWASTE regulations are too stringent by a factor of 1000 and should be relaxed. The procedure does not, however suggest how they be relaxed or whether the cost decrease is large enough to be worth the bother. This is a gap that research should fill, and it seems that this is a task for the Nuclear regulatory Commission.

In the 1980s the ACRS made a study that led to the promulgation of a set of SAFETY GOALS. These were appropriately related to the safety of individuals living near a power plant. But it was recognized that such safety goals were difficult to implement and a subsidiary safety goal was promulgated that the frequency of core melt must be kept to less than 1 in 10,000 years per reactor. Safety improvements must be made to keep core melt frequency below that amount. Although not stated, it was implied that steps to decrease core melt frequency still further were unwarranted and it was not worth the expense to undertake them. Again, of course such a goal is hard to discuss prospectively, but again it can be used as a yardstick with which to compare existing regulatory procedures. For simplicity I will address this "intermediate" safety goal here but the same argument can be applied to the more fundamental safety goal.

A very careful PRA was done by the Nuclear Regulatory Commission in the late 1980s (NRC 1987) for a number of "typical" nuclear power plants. In all cases it was found that this probability was LESS than one in ten thousand per year. There are uncertainties about this calculation, and there has been some discussion about whether one should take the median, the mean or the mode of the probability distribution. I have argued elsewhere that one should take the mean, and do so in what follows. An immediate use of this argument was discussed at a research advisory group meeting. The NRC staff had suggested addition of safety devices to BWR Mark I reactors to improve safety. The committee pointed out that these reactors met the requirement with fling colors. Either the safety goals were wrong, or the NRCs research program that produced NRC 1250 was useless, or the staff suggestion was excessive. In this case the staff suggestion was dropped.

But this seems to be an isolated occasion in which the safety guidelines have been used to limit NRC action. In early 1996 as the result of activities of a disgruntled employee (a whistle blower) it was found that Northeast Utilities had been violating the technical specifications for some years, apparently with the knowledge of a resident NRC inspector. Four large nuclear power plants were shut down because of this violation of procedure. My immediate question, and in my view it should have been the immediate question of the NRC was whether the violation of procedures would bring the calculated core melt frequency high enough to violate the intermediate safety goal or any other safety goal. To the best of my knowledge the answer is NO. The core melt frequency might have been increased by one in a hundred thousand, but the overall core melt frequency would have been less than the goal of one in ten thousand. The actions can be compared to a bus driver driving 35 mph in a 30 mph zone when the road was deserted, with the knowledge and acceptance of the policeman on the beat. It should have taken the NRC no more than 2 months to come up with such an answer. Then the appropriate action might be to allow the reactor to restart either with an insistence that in future the technical specifications would be observed or (even better) that the technical specifications would be modified to allow what seems to be an adequately safe procedure. But the power plants are shut down, permanently in the case of Connecticut Yankee at least. That is comparable to shutting down the whole bus system of New York for the transgression of this hypothetical bus driver.

The effect on public health is large. Supposing the replacement electricity to come from a mixture of fossil fuels and hydro power in the average proportions, each power plant replacement costs over 50 premature deaths a year from air pollution, (Wilson and Spengler 1996) or over 300 deaths so far. Other utility companies have got the message: "Get out of nuclear power as fast as you can". For example in 1997 Central Maine Power has announced that it will not restart Maine Yankee nuclear power plant after a shut down.

This draconian action by the NRC is accompanied by an ever increasing refusal to respond to public comment. When I first complained about what I then thought was inadequate attention to public concern, I wrote, as I always do, to the top man: Dr Glenn Seaborg, then Chairman of the Atomic Energy Commission. Within 3 days he personally telephoned and invited me to Washington to go over my dozen or so issues with his top staff (I did so and spent 2 days in Germantown and Bethesda). In April 1996 I wrote to Dr Shirley Jackson, Chairman of NRC. It was two months before I received a reply. Not from her, but from the Director of Regulation. Even then my query was not addressed. I had asked for technical information. All I got was a general description of procedural violations. After another month of enquiry, I got a partial answer but further enquiry failed to make it complete. It seems that NRC does NOT contest my contention stated above that the increase in core melt frequency is less than one in a hundred thousand per year. I have found no one else who does either. What is going on? Why do we accept this?

If NRC is not following its own guidelines, who should be reminding them of this? When NRC was created, the usual argument was that one should separate the functions of promotion (in DOE) and regulation (in NRC). This argument might have been valid when DOE promoted nuclear power. Now it does not. In spring of 1997 10 Curies of tritiated water were located under the area of the High Flux Beam Reactor at Brookhaven National Laboratory (BNL). Even if someone drank this water for his whole life (which they do not and cannot) his risk would be less than spending a day a week near the radioactive granite of Grand Central Station. Expert hydrologists agree that apart from fixing the leak, little should be done but monitor the slowly moving water in the aquifer and perhaps selective pumping to stop or revere the flow. Yet DOE has taken draconian action that all technical people of my acquaintance believe is unjustified. Not only did Secretary Pena cancel the operating contract of Associated Universities Inc. to operate BNL (which might be justified to maintain public and political confidence) but DOE demand something environmentally incorrect: pumping and aerating the aquifer. Fortunately the dose is minuscule, this does not matter to public health. It only costs money but money taken from our already inadequate research budget. Why has the administration deliberately emasculated the nuclear part of DOE which should have and could have a sane approach to these matters and how can we get it back on track?

DOE is NOT holding NRC to its guidelines and no one else is. Who should? Some people argue that "industry" whoever that is, should be pushing NRC. Yet if a utility urges action on an NRC official, and that official merely delays the operation of a power plant by one day, it costs $1,000,000. Therefore a utility has never shown the will to fight such a powerful regulator. The threat of this much power inhibits the legal action (against NRC) that would be the "normal" outlet in our litigious society.

I suggest that a publicly minded group of scientists, concerned about the three environmental issues that I began with, (global warming, air pollution and sustainability) in reality and not merely, like Vice President Gore and other politicians, for political gain should file suit. Will anyone here join me?

There is a new generation of reactors which are supposedly safer and more reliable. I suggest that there must be a new generation of reduced regulation to go along with the safer reactors. One might for example use "ordinary" pumps and valves in many places instead of the special safety qualified ones which are five times as expensive. The old question of performance regulation versus prescriptive regulation comes to mind. All research that I have seen suggests that in most places performance regulation is better. Yet the examples above and many others show that whenever common sense shows that we must abandon the prescription and merely insist on performance, NRC has dug in their heels. This needs to be changed. I forget who it was who said that "The Power to Regulate is the Power to Destroy". This is happening. Maybe there will be social science research with a multitude of PhD theses on how to regulate the safety of an industry without destroying it. But this will probably happen after it is all over, and the nuclear "industry" is no more.

In 1992 at a meeting on global climate change in Washington DC I made two alternate predictions (Wilson 1994) - the optimistic and the pessimistic. The pessimistic one was that the US would do nothing about this cost increase, and that nuclear power plants would be shut down at an accelerating rate till all were gone by 2017, but that the Nuclear Regulatory Commission would follow a more general law than that of the USA - Parkinson's law (Parkinson 1956). It will still be expanding at 5.75% per year for another 10 years beyond the closing of the last reactor they are set up to regulate. This pessimistic prediction seems to be coming to pass. As a result, the commitment that President Bush made at Rio de Janeiro to keep greenhouse gas emissions to the 1990 level in the year 2000 will not be met, although Dr Allan Bromley made sure that it was a realistic commitment that could have been met.

President Clinton's recent public statements about a climate change treaty, made no mention of nuclear energy. It is noteworthy that he only proposes that the USA reduce to 1990 emission levels beyond his term of office, and even beyond the (possible) two terms of Mr Gore. On Wednesday 22nd October, Charles Babbitt, Secretary of the Interior gave a lecture at Harvard University on the administration's global climate policy. He did not mention nuclear power. In response to a question at a dinner beforehand he explained that the administration believes that the presently proposed goals can be met without nuclear power. He also declined to discuss his refusal to allow the Ward valley low level nuclear waste site to proceed, in spite of the fact that experts agree that the risk is minuscule (well within a reasonable extension of the reactor guidelines above) and California would accept the waste "In their Own Back Yard".

It seems that after the recent discussions with President Jiang of China the USA will allow China to purchase nuclear reactors, and their components, from US suppliers. Let us warn the Chinese NOT to buy our regulatory system along with the reactors unless there is considerable modification.

The overregulation is not confined to the USA. While we all agree that an accident at a nuclear power reactor is bad enough that it must be avoided, that does not apply to subsidiary facilities. In the last 15 years we have been able to prove that US reactors will survive the Design Basis earthquake, and even much worse. But it should not be necessary to design a reprocessing plant with the same extra cost. There is much less inventory of radioactivity and no driving force to evaporate or otherwise spread what there is. Yet the capital (construction) cost of the Thorpe facility in the UK was greatly increased by the (UK) government demand to make it earthquake resistant (Hill 1997). The public furor that surrounded a sodium fire at the Monju plant in 1996 was out of proportion to the effect. That was expected. But the reaction of the Japanese Government was also excessive. So also was the reaction to a small leak of Cs 137 with NO public health consequences in March 1997.

Of course there are cases of inadequate regulation. That usually comes from inadequate alertness. There are also occasiônal "whistleblowers" who for whatever reason raise issues that they feel have been neglected. In such situations there is a temptation for a regulatory authority to tighten up all round in the hope of reassuring the public. It does not reassure me. It suggests that the NRC does not know what it is doing, and merely responds to public pressure. My plea is for a continued detailed discussion of meeting the safety goals and the ALARA principle (which began with the first NRC). Hopefully this could be done in parallel with such a discussion of the safety and environmental goals of other energy technologies which, as the particulate debate shows, has been woefully inadequate.

I suggest that as administration officials go to Kyoto, it will be very important for them to present a research agenda which includes a serious, honest and believable study of why the Rio commitment was not met, and how such commitments can be, and hopefully will be met in future. Reducing the (inflation corrected) cost of nuclear power to 1972 levels is obviously one of the possible actions that the United States could take if we choose to do so. Not to study the obstacles, technical, and political, would be to fail the American people and the world. There is still time for them to become sensible. Let us help them to learn. Finally I appeal to everyone not to be apathetic in this discussion. I know of no proof that people are or will be sensible. I know of no proof that the human race will, or should continue to exist. We must all play our part if we want it to do so.

REFERENCES

Benkhe W. (1997) Communication from W. Benkhe, former CEO of Commonwealth Edison Co. owner and operator of Dresden II and Dresden III

CMP (1972) Report from Central Maine Power, Majority Owner of Maine Yankee. This cost does not include a (later) cost of $20 million to remove a causeway and improve tidal flow in the coolant estuary (which many experts thought was unnecessary and certainly would NOT and have been, and was not demanded of a fossil fuel plant).

Hill, J. (1997) communication to the author by Sir John Hill, Chairman of the UK Atomic Energy

Authority at the time.

Kress, T. (1994) Report to Nuclear Regulatory Commission from the Advisory Committee on Reactor Safeguards

NRC (1987) Nuclear Regulatory Commission report NUREG 1250

NYT (1997) Figures reported in the New York Times in spring 1997.

Parkinson, C.N. (1957) "Parkinson's Law" page 12, Houghton Mifflin, Boston

Tengs, T. et al (1995) "Five hundred life saving interventions and their cost effectiveness" *Risk Analysis* 15:369

Towers and Perrin (1995) report

Webster W. (1972) Letter from William Webster, President of NE Electric System to Richard Wilson

Wilson R. (1994) "The Potential for Nuclear Power" in Global Energy Strategies: Living with Restricted Greenhouse Gas Emissions, edited by J.C. White, Plenum Press, NY. pp. 27-45.

Wilson, R. and Spengler, J.D. Eds. (1996) "Particles in Our Air: Concentrations and Health Effects" Harvard University Press, Cambridge, MA 02138

INDEX

MIX
Papier aus verantwortungsvollen Quellen
Paper from responsible sources
FSC® C105338

If you have any concerns about our products,
you can contact us on
ProductSafety@springernature.com

In case Publisher is established outside the EU,
the EU authorized representative is:
Springer Nature Customer Service Center GmbH
Europaplatz 3, 69115 Heidelberg, Germany

Printed by Libri Plureos GmbH
in Hamburg, Germany